THE LIFE SCIENTIFIC
INVENTORS

THE LIFE SCIENTIFIC
INVENTORS

ANNA BUCKLEY

WITH A FOREWORD BY JIM AL-KHALILI

WEIDENFELD & NICOLSON

First published in Great Britain in 2019 by Weidenfeld & Nicolson
an imprint of The Orion Publishing Group Ltd
Carmelite House, 50 Victoria Embankment
London EC4Y 0DZ

An Hachette UK Company

1 3 5 7 9 10 8 6 4 2

ISBN (Hardback) 978 1 4746 0752 0
ISBN (eBook) 978 1 4746 0753 7

Typeset by Input Data Services Ltd, Somerset

Printed and bound in Great Britain by Clays Ltd, Elcograf S.p.A.

MIX
Paper from
responsible sources
FSC
www.fsc.org   FSC® C104740

www.weidenfeldandnicolson.co.uk
www.orionbooks.co.uk

To my mother, with love

CONTENTS

FOREWORD

This is the second book in this series based on the BBC Radio 4 programme *The Life Scientific*, in which I interview a wide range of scientists and engineers about their lives and work. (The first book in the series was *The Life Scientific: Explorers*.)

The radio series began in 2011 and seems to go from strength to strength, with an average of over 2 million listeners each week, and more who subscribe to the podcast. It is reaching people (men and women, young and old) who thought they had no interest in science, as well as many who already do.

Part of the success of the programme, I am told, is that my guests feel more at ease talking to a fellow scientist about their lives and careers, all the inevitable pitfalls and wrong turns along the way. The conversation flows freely between us. Since the programme is recorded rather than broadcast live, it is then the task of my producer Anna Buckley to edit it down to its 28 minutes for broadcast. More importantly, it is most likely Anna who will have prepared for the show more exhaustively than I ever could. Her very human approach to understanding science, attention to detail and lightness of touch are a winning combination on air and in print.

Everyone in this book can be regarded in some sense as

an 'inventor'. Their personal journeys could not be more varied, and their stories are a joy to read.

Jim Al-Khalili
August 2019

INTRODUCTION

Inventing is about meeting society's needs and fulfilling dreams. But often, it seems, the makers of this world are overlooked. Perhaps because they have no particular desire to join the chattering classes. Or perhaps, as Professor Dame Ann Dowling suggests, it's because engineers 'tend to put their heads down and get on with the job'.

The women and men whose stories are gathered together in this volume have all been interviewed on *The Life Scientific* on BBC Radio 4 presented by Jim Al-Khalili and produced (more often than not) by me. They are the unsung heroes of contemporary Britain: the scientists, engineers and entrepreneurs who have devoted their careers to improving our lives. Some build the infrastructure (digital or concrete) on which we all depend. Others are creating clever devices, smart materials or new techniques. Professor Sir John Gurdon cloned a tadpole. Professor Sir Ian Wilmut and his team created Dolly the sheep. Together, these inventors are moving medicine forwards, transforming the everyday and imagining a more sustainable future for us all.

Many talk humbly of their achievements which typically are huge. A reassuring number had no idea what they wanted to do when they graduated or left school. Naomi Climer 'fell into engineering'. She had wanted to be a cellist. Caroline Hargrove didn't really know what engineering was but was keen to avoid 'accountancy-style maths'. Some think with their hands, others make things in their minds. Many do both. They learn by doing and from books. Making money is often part of the equation too. And there is no space here for the absurd and unhelpful idea that real-world or commercial problems are somehow less intellectually challenging than the abstract things people think about in armchairs or ivory towers.

Engineers working in industry have to find solutions. Ailie MacAdam was dismayed to discover that the 680 historic cast-iron pillars that were supporting the platform at St Pancras station were in danger of 'breaking like carrots'. Professor Lord Robert Mair prevented Big Ben from leaning like the Tower of Pisa, when a new Underground station was built beneath Westminster.

So many of the people I talk to seem to see science as a body of knowledge that needs to be accepted, understood and learnt (a view that perhaps tells us more about how science is taught in schools than anything else). The men and women featured in this volume have a different approach. They solve problems. And if at first they don't succeed, they certainly try again. They challenge conventional wisdom or ignore it. They ask, 'What if . . . ?' Fed up with the air pollution in his home town, Professor Tony Ryan designed clothes that soak up it up. Professor Ann Dowling and her team spent ten years 'on and off' designing a silent aeroplane. Professor Molly Stevens invented an injectable gel to heal badly broken bones.

Some of those scientists and engineers are academic high-flyers, sailing through exams, degrees, PhDs and research jobs. Others take a different route: learning on the job, at evening classes or returning to education later in life. The computer pioneer Stephanie Shirley was 'too tired of being without money' to consider a degree. She wrote code for some of the first business computers in the world (at home while looking after her son) and built a multimillion-pound software company, before most people knew what software was. Professor Mark Lythgoe was too distracted by the Manchester music scene to study when he was at school. He failed most of his A Levels and decided many years later that he wanted, 'more than anything', to do a PhD. He now runs one of the largest medical imaging facilities in the world, pioneering new ways of seeing the human body and treating disease.

These are stories of personal endeavour, full of twists and turns, excitement and frustration. Some of my favourite moments occur where the personal and the professional are intertwined. Adrian Thomas took up paragliding to learn, first hand, how birds fly. Scientists do not live in hermetically sealed laboratories. Engineers have lives too.

Being stabbed in the back when he was a boy introduced Mark Miodownik to the razor-sharp properties of steel. A degree in metallurgy and a PhD on metal alloys followed. Chris Toumazou used to worry that a fuse might blow while his mother was watching *Coronation Street* and became an electrician. Focusing on current (when everyone in Silicon Valley was thinking about voltage), he came up with a radical new way of designing microprocessors and has invented multiple electronic medical devices, including a tiny cochlear chip that enables children who are born deaf to hear. A short research contract with Standard Fireworks lit Jackie Akhavan's fuse and led her to specialise in high explosives and bomb detection, working at the forefront of the war on terror. A visit to

MIT, during a sabbatical year, reassured Wendy Hall that hyperlinks and search engines were worth thinking about after all.

Many of these inventors are professors. But by no means all. A fair few failed miserably at school. At sixteen, John Gurdon was told that 'it would be ridiculous' for him to pursue his interest in science. He went on to win a Nobel Prize.

John Taylor's headmaster described him as 'practically illiterate'. He is, however, remarkably able to see how objects fit together in 3D space. 'You have to run your disadvantages into advantages,' he says. And he feels it is our duty to give something back: 'I am not deeply religious, but I do believe you should leave the world a better place than you found it.' His safety controls are found in almost every kettle in the world.

I am grateful to all these brilliant men and women. Their inventions are changing my world. And their stories are a constant source of energy and inspiration. If one of them were to galvanise you to consider making something new, whether you are at school, retired or stuck on a bus wondering what to do, I would be delighted.

ADRIAN THOMAS

*'There are so many good ideas
out there in nature that we haven't
exploited yet'*

Grew up in: a village south of Oxford
Home life: married to civil engineer Sue Thomas, with a
   daughter, Lauren
Occupation: bio-mechanical engineer
Job title: Professor of Biomechanics at Oxford University
   and Chief Science Officer for Animal Dynamics
Inspiration: watching dragonflies in the garden
Passion: visualising airflows around insect wings
Mission: to understand the mechanics of animal flight and
   design more efficient vehicles
Favourite invention: Skeeter, the dragonfly drone
Advice to young inventors: 'Do something you love'
Date of broadcast: 31 October 2017

7

Professor Adrian Thomas studies how animals fly. He spent almost a decade designing experiments to capture, very precisely, the aerodynamics of dragonfly flight. Inspired by their extraordinary aeronautical abilities, he invented a tiny dragonfly drone with flexible wings which could be used for military surveillance, search and rescue missions and farming. He is also working on a three-legged wheelchair capable of crossing rugged terrain and a boat propelled by a flapping fin.

Jim Al-Khalili interviewed Adrian just before the latest model of his dragonfly drone was about to be subjected to rigorous indoor tests by the Ministry of Defence.

Adrian has spent 5,000 hours hanging in the sky, learning to fly, 'Most of it just for fun . . . Some of it working out new designs for wings.

'Flying in the UK is fantastic,' he told Jim. He goes to one of the Downs and finds a beautiful, grassy, smooth, rounded hill. Then he waits for the wind to come up the hill and launches himself 'exactly like launching a kite', lifting the paraglider above his head while he waits for a thermal updraft to reinforce the wind and pick him up. With some columns of hot air rising at a rate of 5 metres a second, hitching a ride on a thermal is a winning strategy.

Adrian has won the British paragliding championship four times. But nothing beats flying with a feathered friend, gliding and climbing over long distances with the same bird. 'I've had rides where I've climbed out from Milk Hill in the Pewsey Vale and then flown to Swindon,' Adrian said. 'At Swindon I was joined by a red kite which flew with me forty kilometres to Oxford, and then I landed in the field next to my house, which is always rather nice.' Buzzards are good company too. Adrian has become attuned to the distinctive call they make to alert their flock to a rising thermal airflow. 'I listen for that call,' he said, 'see a buzzard circling away, and I'm off straight towards them!'

'Did you always want to fly?' Jim asked.

'I've always been interested in flying,' Adrian replied. 'But

'At Swindon I was joined by a red kite which flew with me forty kilometres to Oxford, and then I landed in the field next to my house, which is always rather nice'

I've also always been interested in the way things flow – aerodynamics, I suppose.' As a boy, he made model aeroplanes and kites. One 'particularly nice' design for a kite was created out of a sheet of A4 paper and a cotton reel and could 'fly unattended for hours'. And he did a lot of sailing, mainly in dinghies. 'I was very interested in how the sails worked together or how you tune them to work with each other.' He enjoyed tweaking the sails to improve the performance of the boat. When the air flowed smoothly over the sails, the boat picked up speed. The faster the airflow, the speedier the boat. But if you tighten the sails too much, you choke off the wind and kill the power. He also thought about the flow of water around and under the bottom of the boat and spent a lot of time trying to design go-fast hulls.

The way air, or water, moves around objects can propel them forward, slow them down or move them from side to side. 'It's really the flow – that's the thing that interests me most.' And it was intellectual curiosity, more than a desire to fly, that first persuaded Adrian to launch himself into the sky. He was studying for a PhD on bird flight and thought the best way to get to grips with aerodynamic forces might be to build himself a set of wings and experience them first-hand. Taking up paragliding also provided welcome respite from some 'hideous [university] politics' which resulted in him being shunted between supervisors and generally feeling unloved. It was a joy to be up in the air, free as a bird, when the power struggles within the university were getting him down.

Later in his career, he mastered another flying machine in order to study the energy savings made by birds when they fly in a V formation. 'I managed to persuade a research council that they should pay for a microlight aircraft,' he said, still slightly amazed that they agreed. 'Wings push the air downwards,' he explained, 'so outside of the wingtips the air is going upwards.' This upwards airflow is an energy source. 'If you're a bird flying behind another bird, you can sit and surf on the updraft on the outside of the wingtips.' Paragliders can do the same, experiencing a bouncy 'ba-dum ba-dum' when they enter the slipstream of another glider. Dinghy sailors, similarly, can slip in behind a faster boat and find themselves planing in its wake. The acceleration is thrilling.

Adrian used his microlight to get into the air with the birds. He chose the biggest he could possibly get and 'went down to the fens to chase V formations of whooper swans . . . I got into the air, dived on the whooper swans out of the Sun – as you do, having read *Biggles* as a kid – and I could see those birds peering over their shoulders and looking at me as I lined up behind them.' He had hoped to sneak up behind the birds and measure the distances between them. He had also planned to film the swans and later count the number of wingbeats per minute. 'But they just accelerated and accelerated and accelerated and cleared off into the distance. They're supposed to do forty knots, maybe fifty at a push. I was doing sixty knots, and they just had extra power there.' This was most unexpected. When squawking and flapping swans take off from rivers, it always looks like such a struggle, so Adrian had assumed that they had very limited available power. As soon as he was up in the air it was clear that this was not the case. High-flying swans, he discovered, accelerate with ease. 'I spent a good week trying to chase them, and they just had the performance edge over me!' he laughed. 'I know when I'm beaten . . . they got me!' He also sensed that his not-so-surreptitious attempts to blend into a wedge of swans, while travelling in a microlight plane, were unwelcome and, not wanting to stress the swans, he decided to stop. 'They're beautiful birds, and they only have a small margin of available energy. I wasn't going to disturb them.' The experiment was cancelled and the microlight was returned. Adrian failed to get any results but lessons had been learnt.

As a professor of bio-mechanics (the physics of how living things move), Adrian combines his two passions: animals and 'the flow of things'. His love of animals was apparent from an early age. 'I grew up in a house with all sorts of animals all over the place,' he said. And, thanks to an ingenious childcare solution, he spent a lot of time with stuffed animals and fossils too. His dad, a senior lecturer in physics at Imperial College in London, would take Adrian and his little brother with him to work and drop them off next door at the Natural History Museum, 'to give [his] mum a break'. (He was one of five children.) 'The two of us would run around the Natural History Museum all day ... I'm not sure you'd get away with that now,' he said. 'I used to walk through those fantastic galleries, looking at the fossils.' The magnificent remains of ichthyosaurs (ancient marine reptiles that lived at the time of the dinosaurs), each spanning several metres, hanging on the walls of the Natural History Museum, never failed to impress. An art report when he was young contained the phrase: 'I wish he would draw something other than sharks and dinosaurs.' 'So nothing unusual there,' Adrian

'I grew up in a house with all sorts of animals all over the place'

said. Slightly more unusual was his well-informed and heartfelt desire, aged six, to be an ichthyologist.

The time Adrian spent exploring the 'weirder recesses' of the Natural History Museum 'definitely had a significant impact', he said. Zoology was perhaps an obvious degree choice for someone who had spent so much time in the company of animals (living and extinct). But when he applied to university Adrian didn't know that it was possible to study such a thing. He went along to his school's careers fair and none of the listed professions immediately jumped out at him. 'The only thing remotely aligned with my interests was medicine,' he said. He applied to Oxford and was interviewed by a group of medics. They asked about his interests and he talked about 'non-medical things'. 'I think we talked about fish quite a lot!' Adrian said, smiling. 'We certainly talked about birds' tails and the aerodynamics of tailless aircraft at some point.'

Detecting where his real passion lay, the interviewers sent Adrian off to the zoology department. There he was handed a deep-sea coral and invited to share his thoughts. 'I was completely fascinated by it,' he said. He chatted away, forgetting that he was being assessed, and later when he was offered a place to read zoology, he thought, 'Wow, I didn't know you could do that!' He is still grateful to the medics for sending him next door. 'I'd have been a useless doctor!' he said, laughing.

During a third-year undergraduate research project, Adrian found his niche: the mechanics of animal flight. He was studying fruit flies and became interested in mutants that had two sets of wings. Curiously, these four-winged fruit flies couldn't fly. They took off and instantly crashed to the ground. And Adrian wanted to know why. Using the kite-making skills he had honed in childhood, he experimented with enhanced design features to see if he could improve the flies' flight performance. Several weeks later, he discovered that a single silk thread tied around the thorax provided much-needed stability, acting like the tail of a kite. The mutant flies were liberated and Adrian started to appreciate the importance of tails.

As anyone who has made paper aeroplanes will know, aerodynamic stability is not easily achieved. The way air flows around a fruit fly, a kite or a paper plane, is notoriously difficult to predict. The same is true for the flow of blood through our veins, or water in streams. Imagine trying to anticipate where and when the next breakaway channel or whirlpool will form in a fast-flowing stream. Newtonian mechanics don't apply and the Navier–Stokes equations (which describe fluid dynamics) are notoriously difficult to solve. If only fluid flows were as simple as balls rolling down slopes.

Having completed his PhD explaining why birds have tails[1] (which took much longer than he had hoped), Adrian knew what he wanted to do. He applied to every animal flight research laboratory in the world and accepted whatever opportunities he was given. A series of very short post-doc jobs followed in Sweden (at Lund, then at Gothenburg), at the Institute of Zoology at London Zoo, in Australia (at James Cook University in Townsville, North Queensland, and later at the University of Brisbane), and at the University of Stirling in Scotland. Moving about like this was a fantastic experience, but 'felt precarious' at the time. 'I didn't make my life as easy as I could have,'

---

[1] Bats don't, and they fly pretty well.

Adrian admitted. But in hindsight he believes 'it was the best thing [he] could have done'. 'I ended up working with all the major figures in the animal flight field at various times,' he said. He also got to see many of the wind tunnels that had been built to study how precisely birds fly. 'I spent all my time – *all my time* – in wind tunnels,' Adrian said.

This simple technology (which involves blasting air through a tunnel) was invented in 1871 to test different designs for flying machines. If you can't move a plane at high speed through the air, why not push fast-moving air at the plane instead? (It's the movement of two objects relative to one another that counts.) Air is driven through an enclosed tube, around a model aeroplane suspended on a wire, and the forces acting on it – lift and drag, for example – are measured using appropriate balances. Testing ideas using scaled-down models placed in wind tunnels is a lot cheaper than building and flying full-blown prototypes.

When the pioneers of early flight, Orville and Wilbur Wright, built a wind tunnel, their flying machines took off. Their previous attempts to simulate flight conditions had involved attaching models of different wing designs to the spokes of an extra wheel that they had fitted onto the handlebars of a bicycle. They then pedalled furiously around the streets of Dayton, Ohio, with this third wheel spinning horizontally in front of them to create an artificial breeze, while simultaneously trying to measure the forces acting on the model. Accuracy was a problem. Suspending a model plane on a wire in a wind tunnel made measuring the forces a lot easier.

Between September and December 1901, the Wright brothers tested two hundred possible wing designs in a wind tunnel and the results of these experiments fed directly into the design of their famous Flyer, the machine that made history in 1903 when it transported Wilbur half a mile through the sky.

Supersonic wind tunnels were built in the 1920s and were used to test possible designs for the new jet planes. Later, animal flight researchers built customised wind tunnels so that they could take detailed measurements of birds as they were flying. Starlings, swallows and house martins were studied in this way.

The physics of flight are the same for a bird or a plane, but swallows are not the same as aeroplanes. The perturbations caused by insects are on a much smaller scale again. But Adrian was not fazed. He returned to Oxford and decided to adapt this technology invented by aeronautical engineers to study dragonflies. 'Yeah, so I built a wind tunnel for insects,' he said, laughing.

'I built a wind tunnel for insects'

By filming the precise position of the dragonflies' wings and the rapidly changing smoke patterns in the tunnel, the airflows around dragonflies in flight could be recorded using a technique known as smoke visualisation. The internal surfaces of his insect wind tunnel needed to be exceptionally smooth, as bumpy surfaces would disturb the flow of air. Adrian spent a lot of the time sanding down the working section of the tunnel. Smoke was generated by rubbing Johnson's baby oil onto heated wires. Placing tiny indentations in the wires, to generate smoke streams half a millimetre apart, so that even minor disturbances to the airflow would be registered, was fiddly work. The tunnel was brightly lit to illuminate the smoke and he used two synchronised high-speed digital cameras, capable of recording 1,000 frames per second.

'Why dragonflies?' Jim asked.

'Dragonflies have this fantastic performance,' Adrian told Jim, becoming animated. 'They glide a lot, but they also can out-accelerate a Formula One car forwards, backwards and sideways if they want to!' This explains why, if you are lucky enough to spot a dragonfly out in your garden during summer, you can never keep your eye on it for very long. And they have incredible stamina, too. Many dragonflies migrate from Britain to Africa. The globe skimmer has the longest migration route of any known insect: it migrates across the Indian Ocean.

Dragonflies are also relatively large, which makes it a little bit easier to see what's going on. (British dragonflies typically have a wingspan of about 10 centimetres.) And, at a rate of twenty wingbeats per second, they flap their wings more slowly than most other flying insects. Conveniently, there were plenty of them living nearby in the Oxford University parks.

Adrian would catch brown and migrant hawkers and ruddy darters in nets in the morning and introduce them to the wind tunnel in the afternoon. (If an overnight stay was required, the dragonflies would be refrigerated to prevent them from damaging their wings.) 'We put them on the perch and they took off and flew around the wind tunnel.' Getting to the other end of the tunnel was not a priority for the dragonflies. They darted about at high speed, and seemed desperate to avoid getting smoke in their eyes.

In a later experiment, Adrian found a way to get bumblebees to fly in a straight line through the tunnel. A beehive at one end of the tunnel and some flowers at the other was all that was needed to get bees to reliably commute through the test section. 'I mean literally commuted. It was like the M25 in the wind tunnel at times,' he laughed. 'It was great fun, really good fun.'

He tried bribing the dragonflies with tasty fruit flies to get them to fly direct, but they were not so easily corralled. So he focused on free flight.

'Most previous studies [of insect flight] had solved this problem by tethering the insects' and measuring the forces acting on them. The insects would frantically flap their wings while scientists dutifully made measurements. But Adrian believed that studying the wingbeats of an insect that is 'trying very hard to get away' is 'no use at all'. A tethered dragonfly has only one thing on its mind. It wants to escape. A dragonfly that is darting about uses its wings in many different ways and is not just flapping to be free. 'You need to know that the animal is doing what it usually does to fly.' For this reason, Adrian was determined to let his dragonflies go where they pleased, even if it did make the experiment much harder to design.

It took about eighteen months to build the wind tunnel and to get everything to work, and just a few hours to collect data. The analysis took a year. But the results were worth waiting for. Existing models of dragonfly flight were proved to be wrong and a range of sophisticated flight strategies was revealed. Dragonflies have four independently controlled wings and direct muscles controlling each of these wings. 'They can flap the front wings and not the hindwings; they can flap the left wings and not the right wings,' Adrian explained. Acting in unison like

a well-trained rowing crew, the wings can all stroke together to achieve maximum acceleration. Or they can
employ counter-strokes to change direction or perform
other manoeuvres. Slow-motion video footage and smoke
patterns showed how dragonflies constantly adapt their
wings to achieve different results and can even change
their strategy within a wingbeat. 'The wings can be halfway through a stroke and they can change the angle of the
wings, or even the shape of the wings!' Adrian exclaimed.
It is this ability which probably accounts for their extraordinary effectiveness as hunters. One study put their
success rate at about 97 per cent.

This extraordinary aeronautical ability is all the more
impressive when you consider all the other things a
dragonfly has to do: eating, growing and reproducing, for
example. It is also worth remembering that these flying
machines were not designed from scratch. Their evolutionary history is relevant too. 'They have sixty to seventy
muscles, depending on species – usually it's sixty-four – in
the thorax,' but not because they need that many muscles
to move a wing. 'They are there because those muscles
started out as the muscles that allowed a worm-like an-
cestor to crawl through the mud.' And allowed some of
their ancestors to be the most ferocious predators in the
pond. As different species of dragonfly evolved over 300
million years, these muscles were then co-opted for a new
function: flying, not crawling.

The results of all of Adrian's experiments were published
in a paper in the *Journal of Experimental Biology* in 2004
and were accompanied by many hours of slow-motion
video footage of the swirling smoke patterns created by
dragonflies flapping their wings, showing how the air
flowed around them. The paper described how dragonflies

manage to achieve such an impressive performance in the air and was of great interest not only to fellow academics but also to the military.

}

A decade after this paper was published, Adrian was approached by the Ministry of Defence and asked if he could make a drone. The British Army were interested in developing lightweight surveillance micro-drones that could be carried by soldiers and released when they wanted to see round corners, for example. (Adrian thinks the same technology could be used to deliver humanitarian aid to war-torn areas or disaster zones.)

'We'd done, ooh, fifteen years of work on insect flight at that point,' Adrian said. 'And I thought, we know enough now to try and build one of these things.' He explained: 'It's a project that's funded by the Defence Science and Technology Lab – Dstl. It's kind of the UK's equivalent of DARPA [the American Defense Advanced Research Projects Agency] . . . part of the Ministry of Defence.'

'Did you ever have any qualms about doing defence-funded research?' Jim asked.

Adrian laughed. 'Turning to the dark side . . . Yeah, absolutely. I was dead set against military-funded work for many years, and so are, still, several of my colleagues. The first set of military funding I got was from the US Air Force, the Air Force Overseas Research Lab. I got that for studying the wingbeat kinematics of hoverflies – so, flapping insect wings. My thought at that point was, well, they could be spending it on bullets, or they could be spending it on me studying insect wings. But the interaction with those guys was really rewarding.' Three sides of A4 explaining the idea was enough to secure $100,000 for three years from the US Air Force. 'They find people who they think are interesting and they fund them.' They monitor progress regularly and stop funding any projects that are 'not delivering the goods'.

Jim wondered if there might be a lesson there for publicly funded research councils. 'Could they adopt a similar idea?'

'Not all is rosy in the defence science area,' Adrian admitted. But the sort of feedback loop they've installed to make sure that people are actually delivering is working well for him. 'It's very much performance-related pay . . . which in some ways can be really good.' Researchers who are dependent on research councils for funding 'can end up spending months and months writing grants. I've got colleagues who do this, who write grant after grant after grant.' Much less is done at the end of these grants to check whether people have actually delivered what they said they would. 'The military guys' do it the other way round. They are happy to splash some cash up front but 'check up on [the projects they have funded] really regularly and decide whether the scientists involved should get paid depending on their performance.' It suits Adrian. 'It's great,' he said. 'So long as you're delivering the performance! That's a hard driver, but why not?'

Adrian is delivering results. The challenge was to design a drone that was much cheaper, smaller and lighter than existing models. 'I don't think I've ever had so much fun,' Adrian told Jim. His plan was not to copy the design of a dragonfly but rather to learn from it. 'I think copying nature is a mistake,' he said. The trick is to understand the underlying principles that make nature's designs so successful at performing a particular task. 'You need to look at what the animal is trying to achieve, and work out the most efficient way of achieving the same target.'

As well as being lightweight, speedy and highly agile, dragonflies are exceptionally good at holding their own in high winds. 'I was out in my garden the other day and it was blowing twenty or thirty knots, and there were dragonflies hunting in the garden,' Adrian said. A tiny flying machine capable of all these things was very much in demand. Most existing drones are bulky and heavy and don't function well in gusty conditions. They are like mini helicopters, powered by propellers, not wings, and they have a terrible tendency to drop out of the sky like lead balloons when their batteries fail. 'YouTube videos of the injuries from quadcopters are horrific!' said Adrian.

Adrian's dragonfly drone, named Skeeter, was a radical redesign. Ever since the eighteenth-century English engineer, George Cayley, made the important observation that flapping wings are not a prerequisite for flight, flying machines had been designed with fixed or spinning wings. Now here was a small flying machine with flexible flapping wings. It has a wingspan that is no longer than a biro and it weighs about the same as two £1 coins, including the camera, communication system and navigation tools it carries on board. It gets off the ground by flapping its mechanical wings and can glide gracefully if the batteries fail, or if a quiet landing is required.

The first challenge was achieving lift-off. 'We solved that one a while ago,' Adrian said. Now it's all about improving the energy efficiency. The team plots the grams-per-watt ratio they have achieved at the end of every week on a wall chart in their office, which shows they are making steady progress. At nearly 12g/W, Skeeter is still some way behind what real dragonflies can achieve (25g/W) but it is well ahead of the competition. The only other lightweight drone, which was inspired by the way hummingbirds fly, can manage just 6g/W. 'We've got a significant lead over what other

> *'We've got a significant lead over what other people have said they can do . . . But you always want more!'*

people have said they can do,' Adrian said, laughing. 'But you always want more!'

Adrian thrives on the synergy between his academic research and working for his spin-out company Animal Dynamics. 'If the company stopped now and I went straight back to just doing the academic thing, I've got more research papers than I have time to write them up . . . so that's fantastic.' But combining being an Oxford professor and an entrepreneur has not been easy. 'I've been juggling away and trying to keep everyone happy and trying to do both jobs properly, and I know it's squeaking in places . . . so it is a cause of stress.'

He said, 'It's difficult, it's weird . . . the university is a shareholder in the company and if the company does well, then the university does well. The university likes generating spin-out companies because it's clear demonstration of impact. But there's not really a good structure in place for academics to combine activities into . . . The standard contract for an Oxford academic allows you thirty days, I think it is, per year for consultancy work. So you have to squeeze it into that. That sort of works, but it's very difficult.'

Jim was sympathetic: 'Especially if your company's just

starting up and you want to put a lot of effort into it.'

'Yeah, exactly,' Adrian replied. 'Start-up companies, they're not a nine-to-five thing. But also you only do them because you really, really want to.'

As a young man Adrian was a paragliding champion, proving himself in the air. Now he is determined to beat the world record for human-powered water travel. Inspired by Skeeter's flexible flapping wings, he wants to build a boat that is powered by a flapping fin. A study by Harvard scientists showed that fins are 20–30 per cent more energy efficient than propeller-based water travel. Fins made from flexible materials also have other benefits. They are kinder to marine life. Fish that get in the way are most likely to be pushed aside, not sliced in two by sharp propeller blades.

In an ideal world, Rolls-Royce or another generous sponsor would provide funding for this radical redesign of ship engines. If they did, he 'certainly wouldn't say no'. For the time being, he is working on a proof of concept and relies on help from a team of retired engineers who still love a challenge. One of his 'silver engineers' has spent several months testing different fin designs in a custom-made water tub that he built in his back garden. The boat they have designed would not look out of place in *Wacky Races* – a souped-up pedalo boat, powered by a flapping fin, not a paddle wheel. In it, they hope to steal the world speed record for human-powered water travel from a team of MIT scientists. The pedal power required is equivalent to cycling up a 1:20 hill at 40km/h for 10 seconds. 'Tough! But not impossible . . .' said Adrian.

Another irresistible project is a three-legged wheelchair

capable of crossing rugged terrain. Wouldn't it be wonderful if wheelchair users could enjoy walking on the moors or hiking on coastal paths? There is no shortage of ideas at Animal Dynamics. Adrian would happily pursue them all at once, his mind darting about like a dragonfly, perhaps. But it's very important, he thinks, to have a serious business partner who will say, 'No. Stop it. Don't touch that! Focus on what's paying the bills.' Work on unfunded projects is allowed on Friday afternoons.

One way or another, however, he is determined to transform vehicle design on land and in the sea, as well as in the sky.

STEPHANIE SHIRLEY

*'It was this crusade for women and
the thrill of writing code . . .'*

Grew up in: Sutton Coldfield, having left Germany on the
Kindertransport
Home life: married to physicist Derek Shirley, with autistic
son Giles (who died aged thirty-five)
Occupation: software engineer
Inspiration: computing pioneer Tommy Flowers, an excel-
lent, non-sexist boss
Passion: writing computer programs
Mission: to offer opportunities in computing to women
and to prove that we could succeed
Favourite invention: software for the black box recorder
for Concorde
Advice to young inventors: 'Look after yourself'
Date of broadcast: 7 April 2015

Dame Stephanie Shirley is a software pioneer and entrepreneur. Born to a Jewish father during the Nazi regime in Germany, she came to England, with her older sister, on the *Kindertransport* train. Her first job, aged eighteen, was as a mathematical clerk at the Post Office Research Station at Dollis Hill, where Colossus, the code-breaking computer, was developed in the Second World War. ERNIE, the Electronic Random Number Indicator Equipment for the new post-war national Premium Bond scheme piqued her interest in computers. Writing code at home, while looking after her son, she set up a company of freelance programmers, employing only mothers (until the Equal Opportunities Act of 1975 made it illegal). At its peak, FT (later Xansa) was worth £500 million and employed 8,500 people. Since then, Stephanie has given away more than £50 million to support people with autism and invested heavily in the Oxford Internet Institute.

'People laughed at me!' Stephanie told Jim. 'I mean, you can't sell software!' Most of the computers on the market in the 1960s had been built with a single purpose in mind, and the programs that were needed to run them were either built-in or provided free of charge. The idea of a multipurpose universal machine was just beginning to sink in.) The chief executives didn't understand what software was and they weren't going to be told, and 'certainly not by a woman!'

But Stephanie never let convention get in the way of what she wanted to do. As a teenager, she walked herself to the boys' school across the road to attend lessons in mathematics and science, dismayed to discover that these subjects were not taught at her girls' grammar school. Her sister went to Oxford, but Stephanie decided against a university education. (The closest she could get to a scientific degree was botany, and she had no interest in plants.) She was also 'very, very tired of being without money'. The loving foster family that had taken Stephanie and her sister in when they arrived on the *Kindertransport* from Germany (when Stephanie was aged five) were 'not destitute or anything like that but [they] did live modestly'. Stephanie 'desperately wanted to start earning and to be independent'.

Aged eighteen, she got a job with the Post Office at Dollis Hill, unaware that it was a hotbed of computer science. The world's first programmable computer had been created at Dollis Hill during the war. Named Colossus on account of its size, it could read and decode encrypted German communications at great speed and is said to have saved thousands of lives. (But the machine was taken to pieces after the war, its designs destroyed, and the man who built it, Tommy Flowers, was sworn to secrecy under the Official Secrets Act.) Knowing nothing about any of this pioneering work, Stephanie chose to work at Dollis Hill because it felt like a nice, secure job with a good pension. A stone engraving that read, 'Research is the door to tomorrow' was the first clue that there might be more to her new workplace than she had imagined.

Her first job was fairly menial. 'I started as a sort of glorified mathematical clerk,' she told Jim, smiling. She would sit and type away at 'one of those great big heavy desk calculators,' following mathematical formulae to get answers to Post Office problems. Keen to progress and encouraged by Tommy Flowers, she enrolled in evening classes after work and completed a degree in mathematics. Using what he'd learnt from Colossus but still sworn to secrecy about his wartime achievements, Tommy Flowers was making a machine that could generate random numbers to select the winners for the new post-war Premium Bond scheme, and Stephanie joined his team. The Electronic Random Number Indicator Equipment Tommy created weighed 2 tonnes and was known as ERNIE. With its flashing lights, *Dr Who*-like noises and regular appearances on TV, ERNIE became something of a celebrity. Investing £1 to possibly win £1,000 was an exciting idea, but people were suspicious of the machine that selected the winners. Some thought they detected a bias in favour of the south-east and one of Stephanie's jobs was to check that the winning numbers selected by

> 'I started as a sort of glorified mathematical clerk'

ERNIE had indeed been chosen at random. They had. There was a more innocent explanation: people living in more affluent parts of the country bought more bonds.

Working with Tommy, Stephanie found her vocation. But the harder she worked and the more ambition she showed, the more she began to feel that she would never receive any kind of promotion. She was the only woman on the team and dressed to fit in, wearing dark suits and pin-striped blouses with a black neck band that resembled a tie. 'There was quite a sexist aura about the scientific world in those days,' she explained.

> 'There was quite a sexist aura about the scientific world in those days'

By day, she was determined to prove to all the men that worked at Dollis Hill that she was their equal and deserved equal pay. When men offered to carry her equipment for her, she would say (somewhat tetchily), 'I believe in equal pay and will carry my own equipment, thank you!' In the evenings, however, she would play down her achievements, fearing it might ruin her chances of finding a mate. Handsome young men would ask, 'What do you do?', and she would say, 'I work for the Post Office', hoping that they would think she sold stamps!

But Stephanie met her husband-to-be at work, not on a random date, and he knew exactly what she was capable of. When they got married, Stephanie decided to look for a job elsewhere, believing it was a bad idea to mix business with marriage. She also quite liked the idea of being able to cash in her £200 pension, a perk left over from the days, not so long before, when married women were required to give up work and were 'rewarded' in this way when they did. She joined Computer Developments

Limited (which later became ICL) in 1960, amazed to find that she could command such a generous salary doing something she enjoyed, and got to work on 'an advanced type of computer' that was being made at General Electric Company in Coventry. Designed to be sold to businesses for a medium price, the ICT 1301 was the first computer to have a core memory and a built-in printer. It weighed 5 tonnes, cost a quarter of a million pounds and was the size of a small fitted kitchen. GEC engineers worked on this vast, hot, noisy electro-mechanical machine during the day and Stephanie visited at night, when the GEC engineers had gone home. Often she would be working in the room reserved for ICT 1301 'pretty much alone'. It was 'this wonderful experience of creativity, of being the first to know something' that made her nights with ICT 1301 so exciting.

Dollis Hill Research Station was 'more prestigious', Stephanie said. 'But the more *fun*, as far as I'm concerned, was Computer Developments Limited. [It] was commercial; it had that drive. We had targets to meet that were much crisper.' There were no targets, however, to promote equal opportunities for women.

Once Stephanie got into business she quickly realised it was a good place for her to be. 'Nobody taught me what you were supposed to do . . . so I just went ahead and did it.' And – following her mantra of going ahead and doing it – Stephanie turned her back on the misogyny at CDL and established a network of women who could code.

'We had no offices at all for the first two years,' Stephanie said. Her coders were all mothers who worked part-time and from home. All Stephanie asked of prospective employees was that they had access to a telephone line. (At that time, most telephone lines were shared, including Stephanie's own and, to this day, she feels guilty about her poor neighbour who must have found the line almost permanently engaged.) She was offering the rare opportunity for mothers to begin their own careers. Since millions of women had given up work when they had children, there were plenty of 'really high flyers' who were keen to use their minds while caring for their children.

Aware of the modern stereotype of coders as young tech wizards, often male, who work in snazzy offices with table-football tables and bean bags, spending hours with their eyes glued to a computer screen and hands typing at almost dangerously fast speeds, Jim asked Stephanie: 'What was coding like back then?'

'In those days it was done with paper and pencil on a desk,' she replied. '[Coding] is a bit like creating a knitting pattern . . . It's, you know, purl one, knit one, cross over, or something like that. And you can follow the pattern and end up with a jumper. Programming is like that. You're giving clear instructions that are repeated at times, but they are absolutely definitive instructions. If you follow [them] you will end up with a pullover.'

'Coding is a bit like creating a knitting pattern'

In coding, 'first, you have to have a method that you can write down in mathematical terms,' Stephanie explained. The precise nature of the challenge needs to be described. What are you trying to optimise? What variables need to be considered? Is there a set of logical steps that would lead a computer to come up with an optimal solution? Manipulating knitting needles according to a particular pattern is routine work. No imagination needed. Working out what instructions would be required to get the jumper of your dreams presents a different challenge.

Once the method is set, the next step is to code it – a process that may take three weeks or three decades. 'Some of the programs that we wrote took years!' Stephanie said, laughing. 'But some of them [took] three days.'

It is a task that is highly creative, infuriating and intensely rewarding in varying proportions. Stephanie almost always found writing code utterly absorbing. She would sit at home and think about all the different ways of approaching the next step, asking herself questions such as: 'What if I turned it upside down and did it the other way around?' Her son Giles was born halfway through her first contract. 'I used to work at my dining-room table with a carrycot at my feet,' Stephanie said. Exploring new possibilities every day and constantly thinking in new directions was, for her, 'a joy'. Some mornings Stephanie would wake up in bed with a solution to a particular task that had been worrying her the day before. Other mornings, she'd wake up realising that she had made a mistake in a program that she'd been working on 'for the last month or so'. The mind works in mysterious ways: 'It's difficult [to know] where innovation comes from ... I wasn't conscious of having made a mistake,' Stephanie said. 'I wasn't churning it over and over.' But clearly, while she was asleep her own biological thinking machine had been running a few checks.

When they had solved the problems they had been set, Stephanie and her homeworkers would post their

'I used to work at my dining-room table with a carrycot at my feet'

34

handwritten instructions to a data centre where they would be turned into punch cards. This early method of data storage was inspired by a system of punched wooden cards invented by Joseph Jacquard[1] in the early 1800s to instruct his new automated weaving looms to either raise or lower the warp. Punch cards stored information as the presence or absence of a hole in a pre-defined position on a piece of stiff card, about the size of an index card. Each card represented one line of code and would, in essence, provide a set of answers to a series of 'yes' or 'no' questions. In this way the card was able to tell the binary computer what to do. A whole stack of cards was needed for these vast electro-mechanical machines, since they had little or no memory of their own. Often the punch cards wouldn't even make it into the computer on the first few goes because something would go wrong with the input routines.

Operational problems notwithstanding, some of the first software packages were made by women working at home, flexibly and part-time, fitting coding in around the needs of their children (and often their husbands too). Stephanie wrote the code that programmed computers to plan the distribution of sugar lorries for Tate & Lyle and schedule timetables for British Rail. Her software was installed in the black box recorders in the new supersonic jet, Concorde.

The early days of the company, however, were not easy going. Stephanie paid herself nothing for several years and had great difficulty attracting new clients. Aware of

---

1 Jacquard created 10,000 wooden cards with holes punched in particular places to create a portrait of himself woven in silk, sowing the seeds of the ICT revolution and demonstrating how vanity can be a highly motivating force.

potential prejudice against homeworkers, she made a tape recording of several people typing and played it whenever she made or received phone calls, keen to create the impression of a busy office and mask any hint of domesticity. She kept writing to business leaders, explaining how much money they would save if a cleverly programmed computer was put in charge of logistics and suggesting they use her company's services, hand-signing each letter with her 'double feminine' forename and surname. No one replied. When her husband suggested switching 'Stephanie' for the family nickname of 'Steve', suddenly responses came flying in. 'Some of them asked to see me, and I would be through that door, shaking hands, before anyone realised that "he" was a she.' Before long business was booming. Initially she wanted to be paid according to the amount of money that her programs saved her clients. Luckily for the clients, they resisted, because often the efficiency savings turned out to be huge.

> *'I would be through that door, shaking hands, before anyone realised that "he" was a she'*

By 1966, Stephanie had seventy-five regular freelancers on her books and an impressive list of clients. Freelance programmers were in demand and there was a real buzz about the business. But the excitement came to an abrupt halt when Giles was diagnosed with autism by Great Ormond Street Hospital, aged three-and-a-half. 'I was basically told that I had to "manage" him better,' Stephanie said, her voice still full of contempt for the way in which she was treated. 'I hate the concept of "managing" my family.' She disliked routine and had been keen to make life as fun as possible for Giles, hoping to delight him with new dishes, messing things up a bit and inventing games. The only time Stephanie fully forgot about work was when she was playing with Giles. 'Of course,' she admitted, 'no mother is perfect, and you look back and think, "What on earth was I doing?"'

The double stress of managing a brand-new company and difficulties with Giles at home did take its toll. 'Both of us finished up in hospital ... Old-style talk would be that I had a nervous breakdown,' Stephanie said calmly. 'Basically, I ceased to function. The depression took over and I was hospitalised, and I just spent weeks weeping hopelessly. My doctor said I should never try to look after Giles again full-time. In fact, he wouldn't let me out of hospital until I had made other arrangements. I think he was probably wise. I might have gone on for another six months and then collapsed again.'

In putting everything into looking after her employees and her son, Stephanie forgot to think about herself. There was no time to focus on the things she wanted to do. 'What I now tell other mothers is, "Look after yourself," and I have learned that you need a healthy selfishness to just survive as a carer.'

Life lesson learnt, Stephanie returned to work. Thirteen years after the company's start-up, the Equal Pay Act was passed in 1975, preventing women from being paid less than their male counterparts. The new law also made it illegal to write job advertisements that exclusively targeted one sex. 'We had to let the men into this women's company,' Stephanie said, 'providing the men were good enough at their jobs!'

Jim commented on the irony that an all-female

organisation, designed specifically to combat sexism and help women in the workplace, found itself in breach of this newly created Sex Discrimination Act. Stephanie graciously acknowledged 'that a mixed workforce is so much more creative and so much healthier, and that's how it should be.'

}

A combination of technical brilliance and her enlightened employment policy made Stephanie Shirley one of the richest women in Great Britain. Having made lots of money in the first half of her career, later in life Stephanie set about giving most of it away. Over the years, she has donated close to £67 million. 'But the more important part of what I've given is ideas and contacts and drive and energy,' she told Jim. Stephanie was one of the first business owners to try out profit-sharing in her sector. She took her company into co-ownership; a quarter of the company went into the hands of the staff at no cost to anyone but her. 'And that, actually, is more important than some of the other gifts I've given.'

Twenty-five per cent of the money donated by Stephanie has gone towards information technology; she put more than £10 million into sponsoring the Oxford Internet Institute's research on the social, economic, legal and ethical aspects of the internet. 'I felt it appropriate to return some of the money to the sector from which my wealth stems. It all came from information technology, so I am glad to give some back.' Stephanie has given about £50 million to help further our understanding of autism and improve the quality of care provided to vulnerable young people with the condition. 'In autism I am the major private donor in the UK, so I am able to have some influence. And I think that's what every philanthropist wants – to make a difference. To leave some marker to say, "I was here."'

Without a doubt, Stephanie Shirley is hugely determined. 'What's driven you to push yourself so hard for so long?' Jim asked.

As a child refugee she understood the need to be resilient from an early age. 'Believe it or not,' she said, 'what [my start on the *Kindertransport* seventy-five years ago] has left me with is as strong today as it was when I was twenty. It left me with the ability to cope with change, to know that tomorrow is going to be different and nothing like yesterday.' Stephanie is deeply grateful to all those who have helped her, and feels she owes it to them to make a difference in the world. 'When your life has been saved and you've had years of people telling you how lucky you've been, you realise, "I don't want to fritter my life away. I want to make each day worth living and my life worth saving."'

> 'You realise, "I don't want to fritter my life away. I want to make each day worth living and my life worth saving"'

'Is it still something that stays with you today,' Jim asked, 'that survival guilt?'

'No,' Stephanie laughed, 'I've had years of therapy to get me out of that! But I certainly had it pretty badly. It's so counter-intuitive. One should be happy to be alive when so many millions died. And yet there one is, weighed down with the responsibility of your own survival.'

Acutely aware of how lucky she was to survive against the odds, Stephanie has taken the responsibility she feels very seriously indeed. It would be hard to accuse her of having frittered her life away. She created a market for software before the word 'software' even existed and, in the space of her lifetime, computers have evolved from massive electro-mechanical installations the size of a small room and yet capable of doing just one thing, to electronic smartphones that can be programmed to carry out an astonishing range of tasks.

In the 1960s Stephanie struggled to convince businesses to let machines do anything more than the most mundane data-processing jobs. Now parents have a hard time getting children to solve problems without them. 'It is unbelievable,' said Stephanie. 'You look back the fifty-odd years and think, "Gosh, isn't it exciting! So much has happened."'

Our ability to think about the world has been transformed and Stephanie is certain it will be transformed again. 'I'm convinced that the next fifty years will be equally exciting,' she said. Who knows what ever-more intelligent machines might achieve? In the meantime let's hope humans might be able to reinvent the wheel and create more opportunities for women who can code.

CHRISTOFER TOUMAZOU

'It was a blessing in disguise, not having that education'

Grew up in: Cheltenham
Home life: married with five children
Occupation: electronic engineer
Job title: Regius Professor of Engineering at Imperial College, London and CEO of DNA Electronics
Inspiration: watching soap operas on TV and worrying that a fuse might blow
Passion: making electronic circuits
Mission: to invent electronic medical devices that help to save lives
Favourite invention: a microchip to detect genetic diseases
Advice to young inventors: 'Be disruptive'
Date of broadcast: 14 October 2014

Professor Christofer Toumazou makes microchips, medical devices and bionic body parts, including the world's first cochlear implant for children who are born deaf, and an artificial pancreas for people with Type 1 diabetes. He left school at sixteen with two CSEs, and trained to be an electrician. Later he became the youngest ever professor at Imperial College London, aged thirty-three, having pioneered a radical new approach to constructing electronic circuits using the 'thumbs approach'. His ultra-low-power analogue microprocessors enabled mobile phones to shrink dramatically in size and made the first digital radios and digital TVs possible.

Jim Al-Khalili interviewed Chris soon after he had won the European Inventor of the Year Award for his DNA microchip, a USB stick which delivers an on-the-spot test for genetic diseases.

'I come from a Greek-Cypriot background,' Chris told Jim, 'and was brought up in a very traditional way, helping relatives who ran tavernas.'

Leaving this 'very molly-coddling environment . . . to go out into this big British wide world made me pretty sensitive,' Chris said. 'I was sort of the black sheep at school.' Having packed lunches that involved the traditional Greek kebabs didn't help. 'All those things that differentiated me [created] this complete sensitivity around me trying to get on with people. I always felt that I was being bullied and I think I was trying really hard to prove myself, particularly in the scientific subjects. I would swot up if I could – with the limited access to books and things that I had – so that I could explain things to [my classmates] the following day . . . I wanted them to think: "Toumazou, oh gosh! He's done all this work. He must be intelligent. He must be brainy!"'

'I would swot up if I could – with the limited access to books and things that I had – so that I could explain things to my classmates the following day'

Despite working hard to stay ahead of the game, he left school at sixteen with two CSEs with medium grades. 'My education wasn't the best,' he said. O levels and A levels weren't offered at his secondary modern school in Cheltenham.

Going into the catering business would have been a 'natural' career choice for Chris. 'Every evening after school my father would pick me up and take me to his taverna where I would be working till about midnight washing up, or helping to serve customers.' But an English uncle made him realise 'there was more to life than a frying pan'. His auntie had broken with Greek-Cypriot tradition by marrying an Englishman, and he introduced Chris to a world beyond restaurants, tailors and barbers' shops. Chris visited the sheet-metal factory where his uncle worked and saw the lathes and the mills. Things that had seemed to have little relevance when he learnt about them at school suddenly seemed exciting. Technical drawings were so much more interesting when he could see how they were being used. His reaction was: 'Engineering – wow!'

Television was another important influence. 'My mother was an avid TV watcher, so we'd all congregate around the TV and we would be watching all the soaps.' While his mother worried what might happen next on *Coronation Street* or in Albert Square, Chris was 'always concerned that a fuse might blow'. Thinking about how TVs worked, he 'got very excited about electronics', despite having 'no formal education whatsoever in that area'.

When he passed his City and Guilds course in radio and electronics 'with a distinction in the electronics part', the family rejoiced. 'It was a really, really big deal for the Greek-Cypriots, a definite *My Big Fat Greek Wedding* moment.'

'Were you then lined up to fix TVs and radios because now you were qualified?' Jim asked.

'I was, I was,' Chris said, smiling. 'The whole of Cheltenham knew I'd got a distinction. Everyone thought I could fix their televisions. But I was very excited about the theoretical side and I knew if I was a technician, I wouldn't be fulfilled . . . so I went on to do an Ordinary National Diploma in engineering, which they say is equivalent to A level.'

He applied to Oxford Polytechnic (now Oxford Brookes University) and was delighted to be invited for an interview. 'My father and my mother, and my uncle and my cousin, and my sister and my brother, and I think another coachload, turned up to Oxford,' Chris said. 'My father even sat in at the interview.'

'The acceptance rate for students with an Ordinary National Diploma was one in several hundred, so I was really lucky to get in.' Once again, his entire extended family celebrated in style but Chris felt 'completely intimidated'. 'This was a completely new world for me. Not just a huge establishment with all these scholars. But forty-five miles away from my family in Cheltenham.' It was too far away to live at home. Near enough, however, for Chris to go home every weekend to help out in the family restaurant.

Chris graduated in 1983 and his teacher, John Lidgey, invited him to stay on to do a PhD. They had done some work together and it had gone well. Chris specialised in 'the thumbs approach', playing with circuits in the lab and having ideas, using the skills he had picked up in his 'vocational-type training'. John applied his knowledge of the theory to work out if the ideas were feasible, or would for ever remain a pipedream.

'[John] introduced me to a completely new playground of electronics,' Chris said. He encouraged Chris to 'do things differently'. And they had a lot of fun, arranging components on circuit boards and trying to get semiconductors to do things they'd never done before.

'It was the antithesis to traditional microchip design'

'The world had been subconsciously persuaded that everything is voltage,' Chris explained. 'When you think power supplies, when you think mains, you think voltage. If you put a voltage across an electrical circuit or a microchip, then a current will flow. And the product of voltage and current is power. I started thinking: what if, rather than putting a voltage across a circuit, why don't we start thinking in terms of current?'

'It was the antithesis, if you like, to traditional microchip design,' an entirely new way of thinking about how to design electronic circuits and make microprocessors. And it delivered some extraordinary results. By focusing on what could be done with the current, the 'voltage levels went down significantly'. In some cases, Chris ended up with circuits that consumed about a hundred times less power than anything else that had been created before.

'That's an incredible achievement!' said Jim.

By thinking differently, Chris invented 'a whole smorgasbord of new analogue circuits', and found himself with a solution that was looking for a problem. He didn't have to look far. 'All these electronic mobile gadgets were being introduced,' and most of them were a lot less mobile than the manufacturers would have liked. (The first commercial mobile phone, the Motorola DynaTAC 8000x, launched in 1983, was the size and weight of a brick.)

'Speech. Sound. Voice. Sight. These are all analogue signals. And quite a chunk of the real estate in these things called mobile phones, and other communicating devices, is actually analogue.' The digital processing bit is only one small part.

In the 1980s, the world was going digital. Rapid advances in digital microprocessing powered a revolution in personal computing. Digital devices were becoming cheaper,[1] faster and more compact, year on year, in accordance with Moore's Law. But 'there was very, very little knowledge of analogue . . . In Silicon Valley in the 1980s, analogue designers or engineers were called dinosaurs,' Chris said. The future was thought to be all about getting logic gates to switch on or off, and processing analogue signals was 'a dying art'.

While most other electronic engineers focused on developing state-of-the-art digital technology, Chris thought about how to process analogue signals more efficiently.

Analogue microprocessors had been neglected by the digital pioneers. They were power hungry and took up valuable space, at a time when everyone wanted electronic devices that were small and lightweight. 'I used the current-mode processing that I'd invented during my PhD to significantly reduce the size of the analogue part of the mobile phone.'

> 'I used the current-mode processing that I'd invented during my PhD to significantly reduce the size of the analogue part of the mobile phone'

The ultra-low-power analogue microprocessors that Chris and John developed were not designed with any particular purpose in mind. They were born out of a desire 'to do things differently'. But there was no shortage of applications. Lower-power processors reduced the need for battery power and, since batteries took up most of

---

[1] The first digital watch, made in 1932, cost about £10,000. By the 1980s they were given away free with breakfast cereals.

the space in many electronic devices, this was a big leap forward.

Chris published papers throughout his PhD, inventing new analogue processors at an astonishing rate. (The external examiner for Chris's PhD said he had written the equivalent of two PhDs in three years.) 'Every time I invented a new circuit, it would go straight to publication,' Chris said. And, invariably, it would get picked up by industry. Everyone wanted electronic devices that were high-speed and low-power. And, more often than not, Chris was able to deliver.

'Slowly I started working with some of these industries and created this concept that applied generically across a number of applications, where there was this necessity to interface the analogue to the digital world.' The world was going digital and radio and TV manufacturers needed help. Panasonic sponsored a lab at Imperial College to encourage Chris to do more research. His current-based microprocessors turned analogue signals into the language of digital electronics. They made digital radio and TV possible. Chris's video amplifier for digital TV was his first commercial invention in the IT sector. It got him into 'entrepreneurial mode' and, together with John, he formed his first small company, LTP Electronics.

It was 'the phase in my life when I wanted to prove I could do everything', Chris said.

His work attracted a lot of funding from industry to set up new labs at Imperial College, but Chris still needed to cover his living costs. He did this by working as a college warden at Holbein House, a hall of residence for Imperial College students. In exchange for looking after 250 male students, he got to live in, 'dare I say, South Kensington'. With Bill Wyman, George Michael and Kylie Minogue for neighbours, it felt like 'the Hollywood of London'. Part of Chris's job was to make sure his charges weren't too socially disruptive. Adopting a strategy of 'If you can't beat them join them', he went through an initiation ceremony to join the Beans Club, 'drinking twelve pints of beer and then swallowing a can of baked beans – not the can, by the way, the baked beans.'

He became a professor when he was just thirty-three years old, an honour more normally reserved for academics in their forties or fifties, making him the youngest person ever to be appointed as a professor at Imperial College, and by quite a margin.

'One day I had an email from a very successful entrepreneur from Canada,' Chris said. His company, Epi Biosonics, had invented an electrode array that enabled children who were born deaf to hear. But it was cumbersome. 'The electrodes were connected to this whacking great digital chip. The digital chip was powered up by something almost as humungous as a car battery. And this whole beast was hanging out of this poor deaf child's head ... He approached me to see if I could convert all this cumbersome processing into a very, very low-power analogue microchip that could be fully implanted in the ear with the electrodes.'

The trick here was to try to mimic the way our bodies process information; to create technology that could do the same job. When Chris was making microprocessors for digital radio, he converted analogue speech into the digital electronics. Here the challenge was the other way round. Chris needed to invent a digital device that could communicate with 'the analogue language of biology'.

(Living cells communicate with one another via tiny electrical impulses that are smooth and continuous. Nature is analogue, not digital.) By adapting the technology that he had invented for digital radio (turning analogue speech into digital pulses) so that it worked in reverse, he invented a cochlear implant that turned digital pulses into smooth, continuous, analogue signals that our brains could understand.

Two years later Chris and his team had made the world's first totally implantable cochlear chip. It is small enough to be implanted in the ear of a deaf-born baby and needs a million times less power to run than a domestic light bulb. The battery is tiny and can be recharged just by rubbing the back of the ear. Tens of thousands of children who were born deaf have had their lives transformed by Chris's cochlear chips.

The best innovation, Chris believes, comes from bringing people with very different skills and interests together – even if, left to their own devices, they might not normally choose to talk to each other. It's hard work but the prize is worth it. 'I love the way I can get engineers and medics and physicists and biochemists to work together,' he said. It's important to allow people to escape the silos that traditionally confine them and to encourage scientists with very different skills and knowledge to work together. It can be a challenge. 'All scientists, and particularly doctors, have a tendency to go off into little corners to talk about medicine together,' one doctor on the team admitted. It takes a while for engineers to understand what the medics are saying and vice versa, but when they do, the conversations that follow are often very exciting.

By the time he was thirty-five, Chris had an impressive track record of innovation in both consumer electronics and healthcare thanks in large part, he believes, to the multidisciplinary 'playground' he had created. Next, he decided to try to make an artificial pancreas to help people with Type 1 diabetes who are unable to process insulin. If Chris and the team could make a bionic ear, then maybe he could create an electronic pancreas. The beta cells are the insulin-generating cells in the pancreas. They measure the amount of sugar and release an appropriate amount of insulin. So Chris suggested to his team: 'Why don't we just take the pancreas and see if we could make a silicon chip that mimics the beta cells?'

'You're not the only ones who are working on artificial pancreases,' Jim said. 'So what's different about your device?'

'The key difference is that this is bio-inspired. In other words, because we're using low-power analogue technology, we're able to mimic exactly the biological behaviour of the pancreas in a way that biology understands.'

When Jim interviewed Chris, his artificial pancreas was in the second year of clinical trials. 'At the moment we're on about sixty, seventy patients,' he said. 'The next stage is for these patients to go home and use these sorts of technology themselves.'

Most of the time, when he's inventing, 'the idea comes before the problem ... It's one of those back-of-the-envelope "What if?"s,' Chris said. 'My PhD students think, "Well, because Professor Toumazou has said, 'What if?', it must be solvable." But [most of the time] I really don't know if it is.' On one occasion, however, the problem came before the solution. Chris and his wife were caring for their son, Marcus, who had lost his kidneys through a genetic predisposition when he was nine years old. 'The poor kid was on home dialysis for three years,' Chris told Jim. It was a very worrying time. They had to keep a close eye on Marcus 24/7 and were constantly worried that his condition might deteriorate suddenly and they would fail to spot the signs. Chronic disease management seemed desperately primitive and Chris wanted to take away some of the stress and anxiety around caring for his son: 'Wouldn't it be nice if we had a big brother, if you like, monitoring Marcus?' What if patients who were at home could be monitored remotely and clinicians could keep an eye on them?

'So, I invented this small processor that can measure things like heart rate, temperature . . . and the processor was embedded in something that looks like a sticking plaster.' These 'digital plasters', as they came to be known, could be attached to the patient's chest, and the processor connected wirelessly, either via Bluetooth or by GSM (the cellular network used by mobile phones), to the hospital IT system. These patient-monitoring patches would take readings continuously (updating every few minutes, not every four hours). The system could be set up so that if the readings received from the patches fell outside a pre-set range, the hospital would be alerted immediately and, if necessary, the patient could be taken into hospital without further delay. By having an automatic early-warning system in place, some of the worry would be taken away: 'You're not constantly thinking, "What's happening? Should I go and check?"'

It could also help to save lives. A study in the *British Medical Journal* estimated that one third of all the preventable deaths in 2012 could have been avoided by good clinical monitoring, taking regular measurements of basic health indicators and spotting when readings fell outside the normal range. Chris's 'digital plasters' could help to save lives. And, if it meant more patients could safely be looked after at home, it might also help to reduce the demand for beds in hospital wards.

Chris's Sensium pad was approved by the US Food and Drug Administration in 2011 and trialled by a hospital in Brighton run by the private healthcare company Spire in 2014. 'Several hospitals in the US and in the UK are now using this technology,' Chris said. 'It's really starting to pick up.'

}

Their son inspired another of Chris's innovations. The family had no idea that Marcus was ill until, one day out of the blue, he suddenly collapsed. Only then did they discover that he had a rare genetic disease that led to chronic kidney failure. This made Chris think: if only there was a simple way of detecting genetic diseases like this. If people knew that they had a genetic predisposition to certain diseases, then perhaps they could take appropriate action to reduce the risks of organ failure. Within months Chris and one of his PhD students had invented a microchip that could search for a specific genetic mutation in a sample of saliva and deliver the results in minutes.

'On this microchip, I've got the template of a particular fragment of DNA that I'm trying to detect. We all differ by [about] 0.1 per cent, and it's this 0.1 per cent difference – all these variants that determine whether you've got predispositions to diseases like my son's renal disease, whether you can metabolise drugs differently, for example,' Chris explained.

'And how exactly do you use a microchip to extract information about a person's genes?' Jim asked.

'Saliva is taken and within that saliva there are fragmental molecules of DNA. Within minutes I can extract your DNA and determine whether or not it matches the template that's on this chip. If it matches, it switches on the microchip.' A current flows and the user knows that the fragment of DNA that the chip was designed to look for is present in this bit of spit.

> 'Within minutes I can extract your DNA and determine whether or not it matches the template that's on this chip'

'So, the test looks at a specific gene in a person's DNA. That affects how well they process a particular drug, so the doctor can then tailor the dose to their needs, for example?' Jim said.

'Absolutely. By personalising medicine you're not over- or under-dosing. [Patients] won't come back into hospital as many times, so the saving to the healthcare system is enormous ... The big difference here is, we've taken a lab and put it on to a chip. And it's the speed of the result that's so, so important.'

The DNA microchip can be put inside a USB stick. A GP can put the stick into their computer and deliver the results of the test to the patient in under 20 minutes.

Multidisciplinary research and spin-off companies are encouraged in academia now, but when Chris's company,

LTP Electronics, first started to do well, not everyone in academia was supportive. Sensing the disapproval of some of his colleagues, some of Chris's childhood insecurities reared their ugly heads. He felt like the black sheep once again.

The arrival of a new rector at Imperial in 2001, however, changed everything for Chris. The new rector, Richard Sykes, was a former chairman of the multinational pharmaceutical company GlaxoSmithKline. 'Richard has been a real mentor for me,' Chris said. 'For him it was all about, "What the hell are we doing with all these separate departments? We need to knock them together . . . That's where all the big science is: environment, healthcare, energy. Let's just make these things happen. Let's get people to work together." So you can imagine Richard and I really hit it off! And that's how our sort of honeymoon together in academia began.'

Together they came up with plans for a new Institute of Bio-medical Engineering at Imperial College, designed to bring doctors, engineers and scientists under one roof to create bio-inspired technology. 'To do that we needed to raise twenty million pounds,' Richard said. 'But Chris wasn't fazed by that. He just said: "No problem! I can go out and raise that, but the university has to do something."' The university put down £10 million and Chris raised the other £10 million – relying, Richard said, on 'his charm, his enthusiasm and his credibility'.

'What do you think has been the secret of your success as an inventor?' Jim asked.

'I feel that I was able to invent things through my intuitive approach,' Chris said. 'It was a blessing in disguise not having that education.' His training enabled Chris to maintain an open mind, to think differently from well-educated graduates. And the way he responded to feeling like a black sheep when he was at school continues to spur him on. 'That challenge to achieve and to really try and be the best, has driven me,' Chris said. 'And I will continue. I see major challenges ahead of me and it's never going to stop.'

JACKIE AKHAVAN

'A contract with Standard Fireworks got me on the journey into explosives'

Grew up in: Peckham, South London
Home life: was married to Shahriar Akhavan, now widowed
Occupation: chemist
Job title: Professor of Explosive Chemistry
Inspiration: her dad, an electrician who made science seem relevant and exciting
Passion: explosive chemistry
Mission: to make the world a safer place for British Army personnel and civilians
Favourite invention: 'the most exciting ones have to be kept secret'
Advice to young scientists: 'Science isn't difficult, but you do have to enjoy it'
Date of broadcast: 30 September 2014

Jackie Akhavan is a professor of explosive chemistry who is involved in the war on terror. She left school not having a clue what she wanted to do and never imagined that she would end up becoming an expert in high explosives, studying Semtex, and designing new materials for warheads. Working at the Centre for Defence Chemistry at Cranfield University, her clients include the Ministry of Defence, the Atomic Weapons Establishment, BAE Systems and the Defence Science and Technology Laboratory (Dstl). Making bombs safer is not a contradiction in terms, she says. She has created shock-absorbent polymers that reduce the risk of accidental detonation. And she does a lot of work on bomb detection too: creating polymers that change colour when they come into contact with known explosives, for example; and testing some more unusual ideas such as training honey bees to sniff bombs.

'Going to grammar school set me up for life,' Jackie said, when describing her childhood to Jim. She grew up in a council house in Peckham in South London and went to Honor Oak Girls' Grammar School in Forest Hill. Her dad, an electrical engineer who fixed the lifts in the Savoy Hotel, taught her everything she needed to know about electric motors in an afternoon, when she was struggling with physics A level. He was less keen, however, on the idea of his daughter going away to university, fearing (quite reasonably, as things turned out) that if she left home at eighteen, she would never return. But Jackie was keen and her mother, who worked at the Department of Health and Social Security office in Peckham, was supportive. So, she went to Southampton University to study science. 'I can say that in the first three weeks the maths was way beyond me, and the physics I couldn't understand,' Jackie said. 'But in chemistry, at least, I could understand what was going on.'

She graduated from Southampton with a degree in chemistry and for a year worked as a technical sales representative selling synthetic latex to companies that made condoms, carpet underlay and other products made of rubber, until a chance meeting with some old university friends, who seemed to be very happily doing PhDs, persuaded her to return to academia. Professor Patrick Hendra welcomed Jackie back to the chemistry department with open arms and ICI were happy to pay for her to study how polymers age, in order to make products that were more robust. Why do plastics degrade and crack?

By the time Jackie had completed her PhD, she knew

she wanted to be an academic. Professor Hendra advised her to get a job in the wider world first, and then come back in, so she did just that, working for three years in the research and development department at Pirelli General Cable Works. Fibre-optic cables made from glass enabled vast quantities of data to be transported at high speed, but they were expensive to manufacture. Jackie's job was to create a clear plastic polymer that could replace the glass in fibre-optic cables. The work was rewarding. Fibre-optic cables made with clear plastic do not perform quite as well as those made with glass, but they work over short distances (between computers, for example) and are still widely used today. Jackie quickly became the company expert on plastic fibre optics and was well respected. But she missed being able to bounce ideas around with colleagues. She wanted to be able to ask questions as well as always being expected to have the answer.

At this stage in her career, she had no desire to work with high explosives. It was when she was looking for a job for a friend that she spotted an advertisement for a polymer chemist and thought, 'That's me!' Stressing her industrial experience working with polymers, and not worrying too much about the small print which said, 'needs to know about explosives', she applied to the Cranfield University at RMCS, not knowing then that RMCS stood for the Royal *Military* College of Science. (It's now known as the Defence Academy of the United Kingdom.) Six men interviewed her. One asked, rather aggressively, she thought: 'What is the point of polymer research?' Jackie defended her field of expertise and decided that she didn't want the job if people were going to treat her like that. When they asked her what salary she wanted, she doubled the figure she had in mind to put them off. It didn't work. To her great surprise and initial displeasure, she was offered the position and, after a few friendly phone calls from Cranfield University, she accepted.

Seeing so many people in military uniform was unsettling at first. The college is based on a secure site at Shrivenham surrounded by a high-wire fence. But the academic freedom was wonderful and the research facilities were amazing. Teaching army undergraduates was also a joy. Not all the army officers were so well behaved. Outside of the classroom, Jackie insisted that they said 'please', and even lieutenant colonels, who were more used to ordering people what to do, usually obliged.

There were plenty of opportunities to do research too. 'A contract with Standard Fireworks got me on the journey into explosives,' Jackie said. At that time the company was staffed by women who made fireworks by hand, putting powders into cardboard tubes. Getting the right mix of explosive powders was a skill that many of the workers had learnt years ago and could repeat almost in their sleep. Jackie had been brought in to automate the process and turn making fireworks from a craft into a science. First she established reliable recipes formalising the precise ratios of the different powders – fuel, oxidiser and metals – that were needed to create pleasing combinations of colour, light, sparks and sounds. She then added polymers to the mix to hold the different powders in place. In this way long cylinders of solid explosives were created that could be cut into pieces by a machine to make individual components for fireworks.

'A contract with Standard Fireworks got me on the journey into explosives'

'Tell me, Jackie,' said Jim, 'what are the ingredients for a really good firework?'

Jackie took a rocket as her example. 'You generally have something to get it up into the air, which is gunpowder.' Despite its brainy reputation, rocket science (at least for fireworks) is relatively straightforward. A tightly stacked stash of gunpowder thrusts the rocket into space when ignited. Gunpowder is made of carbon, potassium nitrate[1] and sulphur. Carbon is the fuel, 'so that's got the *oomph*'; the potassium acts as the oxidiser. The sulphur is the binding agent. It holds the explosive ingredients together and keeps them apart prior to the planned explosion. When the gunpowder is heated, the sulphur, which has the lowest melting point, becomes a liquid and the carbon and potassium nitrate are brought together. The chemical reaction that takes place releases gases that are forced downwards out of the tube, and an equal and opposite upward thrust is created that launches the firework from a standing start to travelling upwards at high speed.

The 'pretty, pretty, bang, bang' displays, as Jackie sometimes likes to call them, are created by a series of much smaller explosions that take place once the rocket is airborne. The ingredients for each of these smaller explosions are kept in separate compartments and ignited in rapid sequence by a slow-burning time-delay fuse. Once the rocket is up in the sky, 'a bit more gunpowder breaks apart all [these] little capsules' and a collection of colourful chemical reactions takes place. Different metal compounds give out light in the visible spectrum when heated to the high temperatures generated by these

---

[1] Also used in fertilisers, and for tree-stump removal, and to preserve pork. Potassium nitrate additives make ham and bacon pink.

explosions. Sodium generates a yellow or orange flame. Strontium glows red, and barium green. Chemistry makes fireworks colourful.

'A chemist friend of mine likes to make his own fireworks from scratch, so it's clearly not too difficult,' teased Jim.

Jackie was quick to stop him. 'Let me tell you, it is illegal for you to make fireworks,' she said, sternly. 'And I wouldn't tell anybody how to make them except for the people who work with me and who are experts.' If anyone has the authority to advise on the safety of handling explosives, it's Jackie.

Once she had mastered fireworks, Jackie graduated to more dangerous explosives. 'Energetic compositions', as the chemists at Cranfield call them, can be classed into three categories: pyrotechnics, propellants and high explosives. Pyrotechnics, like fireworks, burn the slowest, often creating an impressive visual effect. Propellants burn faster and produce a gas that pushes something forward, 'like pushing a bullet through a barrel or lifting a missile off the ground.' High explosives are 'the ones that detonate and the ones that cause all the fragments', Jackie said. Before long this was Jackie's main area of expertise.

The total energy released by even the most devastating explosion is not as much as you might think, she explained. The speed of the reactions is what gives it its explosive power. The scale of the explosion is determined by the burn rate rather than the sheer amount of energy released. 'A

'A candle gives out more energy than an explosive over one day'

candle gives out more energy than an explosive over one day.' The difference is, a candle gives out its energy very slowly. High explosives detonate in microseconds. Temperatures soar to several thousand degrees Celsius and the pressure rapidly becomes intense. This rapid change in pressure generates shock waves, and supersonic shock waves are what cause devastation.

Jackie's first job for the Ministry of Defence was working on a European contract for the Defence and Evaluation Research Academy to improve the safety of warheads. Cranfield University's Centre for Defence Chemistry is the only university department in the UK that is licensed to manufacture and store up to half a kilogram of high explosives. And so Jackie was free to perform experiments with explosive ingredients that would be illegal elsewhere, making and testing a series of tiny bombs. Could high explosives be made safer to handle without compromising their performance?

The challenge for Jackie was to develop a polymer that could be added to high explosives to make them less sensitive to 'adventitious stimuli'. If, for example, munitions were dropped or jostled around in the back of a truck, there was a risk that traditional explosives would be set off by the simple act of knocking into another solid object. But if the explosive ingredients were bound together by a shock-absorbent polymer, then the risk of premature detonation would be greatly reduced. By creating a rubbery matrix, she managed to desensitise explosive ingredients to unexpected shocks without impeding the reaction that would occur when the warhead was ignited.

Jim asked Jackie about the ethical considerations of

conducting research to be applied to warheads. 'We're trying to make them safer for the manufacturer and also safer for the soldiers who use them,' Jackie said.

Nonetheless, isn't making safer explosives an oxymoron? 'Surely, if a bomb's safe, it's not going to go off,' Jim said.

'If you drop it, you don't want it to go off,' Jackie clarified. 'You only want it to go off when it's actually initiated with a detonator.'

'But so long as scientists continue to develop explosives, we will always live in a world where they cause damage,' Jim replied.

'That's very, very true,' Jackie agreed. 'But in life, really, we're human – we're never satisfied with what we've got, and we're always striving for more.' Jackie believes that war is something that will always be around. The best one can do is to ensure the weapons being used are as safe as possible. Jackie works on the research side, and so never gets to see the end result of what she does. 'We're literally innovating new explosives and new compositions for probably ten to fifteen years' time,' she said. 'We never actually see the finished article, but we do feel that we are helping our soldiers to keep safe.'

> 'We do feel that we are helping our soldiers to keep safe'

Jackie's early research was about making existing weapons safer. More recently she has done a lot of work on bomb detection: inventing methods to help security professionals keep pace with the proliferation of improvised explosive devices manufactured by terrorist organisations around the world. During the Troubles in Northern Ireland in the 1970s, 19,000 IEDs were detonated in thirty-five years. That's an average of one explosion every 17 hours. And the significant security threat posed by unauthorised explosive devices hasn't gone away. IEDs are thought to have been responsible for two-thirds of all the Coalition casualties during the recent war in Afghanistan. But ways of stopping improvised explosions before they go off have become more and more refined.

Jackie has used a range of techniques to fight the war on terror, including Raman spectroscopy, the diagnostic technology she mastered for her PhD. This technique measures the ways in which molecules vibrate in order to identify them. Jackie's first attempt to study the chemical properties of explosives in this way didn't quite go to plan. She had contained her sample in a little glass bottle but made the mistake of keeping its top on. The laser heated up the propellant and it degraded, producing a gas that 'shot the top off!' The lid flew open and the sample disappeared in a puff of smoke. Fortunately, the sample was very small and the explosion didn't bring any harm to the instrument (or to Jackie). But it was all a bit alarming.

Safety lesson duly learnt, she moved on to studying Semtex using Raman spectroscopy. Semtex was popular with terrorists precisely because it couldn't be detected but, using some of the tricks she had learnt during her PhD, Jackie and Professor Hendra showed that it *was* possible to identify Semtex. Delighted, they published their results and soon found themselves being asked lots of questions by the UK Security Services, who were not best pleased to learn that two UK scientists were so re-laxed about giving away a new secret counter-terrorism weapon within weeks of it having been acquired. They didn't want terrorist organisations catching on to this game-changing method of Semtex identification.

It was an early lesson in what can and cannot be published when you are an explosives chemist. 'It must be frustrating that some of your work will never get published,' Jim suggested, aware that most scientists want as many people as possible to benefit from their insights, and that the number of papers a scientist has published is often taken as shorthand for academic success. 'We always say that there's a way to make the science publishable,' Jackie replied. 'It's the application that we can't publish.'

Most known solid explosives can now be detected by a simple swab. But the threat posed by a new type of IED was made horribly clear in August 2006 when an al-Qaeda plot to blow up a transatlantic flight was uncovered just two weeks before the flight was due to take off. If everything had gone according to the al-Qaeda plan, suicide bombers would have boarded the plane carrying a rosewater bottle filled with liquid explosives. These prettily perfumed bottles would have sailed through airport security. When MI5 raided a flat in East London where these explosives were being made, they found equipment that was utterly unfamiliar. These bombs would explode when two liquids mixed together, with no wires required to initiate the process.

Airports in Europe and the USA immediately introduced a ban on passengers carrying more than 50 millilitres of liquids in their hand luggage, to minimise the risk to passengers; and chemists, like Jackie, were pulled in to help with bomb-detection work. Up until then most bomb-detection devices had focused on screening for electronics. Airport security, baggage scanners, body scans and frisking are good at detecting wires. And all known solid explosives can be detected from a simple swab, taken from suspicious items. Now, suddenly, understanding how the liquid explosives might work was of utmost importance. The liquids that al-Qaeda planned to use were never identified and there was no guarantee that such a bomb would have worked. But it suggested a radically new approach to making bombs.

'It's a struggle to stay one step ahead,' Jackie admitted. 'The chemists are so, so clever . . . We've got respect for them, because we try to stay one step ahead but, actually, we're probably equal to them. We're learning all the time.'

Several companies made machines to test for liquid explosives. They then asked the Centre for Defence Chemistry to supply them with explosives so that they could check if their machines would do the job. Jackie would say: 'What do you want?' and they would reply, 'We were hoping you would tell us.' It was a hopeless case of chicken and egg.

In the meantime Jackie did her own research. Using her knowledge of polymers, she and her team invented a new material (a synthetic polymer) that would change colour when it came in contact with the vapours released by liquid explosives, or illicit drugs. Detection devices have been made by putting these polymers on to smart chips, and a commercial product called CRIM-TRACK is being developed.

Raman spectroscopy can also be used to analyse suspicious liquids and look for tell-tale chemical signs of potentially explosive compositions in vapours. Ironically, the difficulty arises not because the tell-tale ingredients are particularly difficult to detect but rather because they are so ubiquitous. Scents like Chanel No. 5 and other powerful perfumes generate a Raman profile strangely similar to those of liquid explosives, for example. One way of overcoming this problem could be to run several tests and only treat objects that test positive on all fronts as suspicious. It is a work in progress. The dream explosives-detecting machine would be able to run multiple tests in a short space of time and enable us all to board the plane on time. If this could be achieved, then we would all be able to go through airports with our bottles of water and not have them whisked away.

'There's an array of detectors that will go into airports … And I think in one or two years' time we should be able to detect liquid explosives or the vapours coming from liquid explosives,' Jackie said. 'Hopefully we'll crack that one.'[2] Meantime, if you do get stopped by airport security and are baffled as to why, bear in mind that your favourite fragrance might be responsible for making security personnel view you in a very different light. The chemical profiles of a sweet-smelling passenger and a suicide bomber are similar.

> 'In one or two years' time we should be able to detect liquid explosives'

The need to try and keep one step ahead of the terrorists has forced scientists working for the defence industry to get creative. If sniffer dogs have long been trained to detect bombs as well as drugs, they thought, perhaps honey bees (another animal with an exceptional sense of smell) could do the same? Bees had already been

---

2 They are still working on it.

trained to detect heroin and cocaine and in 2006, scientists at Los Alamos announced that bees could also be trained to detect explosives. Each time a bee detected an explosive and followed it downward, it was rewarded with sugar water. In just a few hours, most bees learnt to stick out their proboscises, as if they had found nectar, when they came across explosives. Machines were being made that, it was claimed, could train 2,000 bees a day and that cost next to nothing to run. Not all the bees acquired this new skill, but those that didn't make the grade were automatically excluded and so a steady supply of reliable bomb-sniffing bees was guaranteed. Trainee bees were easy to find and proved to be quicker learners than dogs. 'I was told that you can train a bee to detect explosives in five hours,' Jackie said. Training a sniffer dog takes three to four weeks and regular refresher courses are needed if dogs are to maintain this skill.

Keen to capitalise on the Los Alamos discovery, a company approached Jackie and her team for help in creating a new bee-based detection device. First Jackie wanted to establish just how sensitive these bees were to the whiff of explosives. She set up a series of experiments to test the bees' olfactory skills by exposing them to explosives in different concentrations. The bees, however, seemed unable to detect any of the explosives. This was very unexpected and disappointing. Perplexed, they investigated further, repeating their experiments and increasing the concentration of explosive, just in case that was the problem, but still their results seemed to contradict what others had discovered. Eventually they realised that the bees were exquisitely sensitive to the additives found in most explosives (used to improve and preserve them) and woefully unresponsive in the presence of the pure explosive that

Jackie and the team had manufactured. 'So unfortunately, the experiment failed,' Jackie told Jim. It proved that using additive-sensitive bees was not a reliable method of bomb detection. Plans to install bee boxes in all major airports were put on hold. Training sniffer bees to work alongside airport security is just one of many ideas that have been invented or tested by Jackie and her team. A rare damp squib in an otherwise explosive career. Jackie is committed to the war on terror. And, for as long as there are bombs in the world, she is determined to use her knowledge of chemistry to keep a step ahead of terrorists who are making IEDs and make explosive devices that are safe to handle.

JOHN C. TAYLOR

*'You have to find a way to overcome
your disadvantages and turn them
into advantages'*

Childhood: evacuated to Ontario, Canada as a boy and
later educated on the Isle of Man
Home life: divorced with three children
Occupation: inventor and entrepreneur
Inspiration: 'I actively didn't want to be an inventor'
Passion: wanting to know how things are manufactured
Mission: 'to leave the world a better place than I found it'
Favourite invention: a tiny thermostatic control for small
electric motors
Advice to young inventors: 'Avoid financial services'
Date of broadcast: 25 May 2018

John Crawshaw Taylor specialises in making thermostatic controls and has over four hundred patents to his name. His integrated thermostatic control system for electric kettles has been installed in 2 billion appliances worldwide. And we have John to thank for cordless kettles and limescale filters. His first commercially successful invention was a safety control for windscreen wipers. Invented in 1965, this tiny device, known as the Otter G control, has sold, on average, a quarter of a million units a weeks for more than fifty years. It helps to prevent fires starting in the small electric motors found in cars, tumble dryers and other domestic appliances. His tiny, inexpensive inventions have made millions and in recent years he has donated more than £10 million to Cambridge University and the Royal Academy of Engineering in London, to encourage innovation.

John Taylor arrived at Broadcasting House smartly dressed in a blazer and cravat, having flown his own turbo-prop plane from the Isle of Man that morning.

'Every time I boil a kettle, the chances are that I'll be relying on one of your inventions,' said Jim. 'Do *you* feel a twinge of pride every time you make a cup of tea?'

'Yes. But I feel even more proud when I go somewhere which is really the back end of the back end, and there in the shop window is a kettle. And I look at it and I think: "Yeah. That's one of mine, that is."'

A small device designed by John is hidden inside 2 billion kettles around the world. It tells the kettle to switch off when steam emerges from its spout and prevents the element from heating up if there is no water in the jug. And if either of these controls fails, an ingenious thermal fuse will make sure the kettle itself doesn't overheat and melt on to the kitchen sideboard.

Modern kettles are much cleverer than copper-bottomed vessels that are put on the hob. And much of their cleverness has come about courtesy of a series of tiny devices invented by John. 'Many of my things are hidden away and never seen by the public,' he said. But his company, Strix, cornered 75 per cent of the global market for safety controls in electric kettles.

As a child John was 'always making things': model aircraft and little tanks made out of cotton reels and rubber bands. 'I used to knit dish-cloths for my maiden aunts as

Christmas presents,' he said. 'My first actual invention, although it wasn't patented, was a method of attaching a little glider on to a kite string ... The wind blew the little glider up the kite string and I had a device like a mouse trap at the top which released it.' The glider would then float down. And the mouse trap would fall down to earth, ready to be used again. 'It was the best small-boy exercising machine' and kept John entertained for hours.

'How did you get on at school?' Jim asked.

'Badly. I'm dyslexic and, of course, that was completely unknown when I was a child, so I was at the bottom of the form in everything except maths or science. And yes, I found it very difficult. I took and failed the eleven-plus exam. I took and failed my thirteen-plus. I took and failed the common entrance exam. Almost in desperation, my father and mother sent me to King William's College on the Isle of Man to take their entrance exam.' John performed relatively well in the maths exam (the only questions he got wrong were the ones that used words to describe the problem), but failed all the exams that required him to read and write. 'The headmaster said I was practically illiterate,' he said. 'I couldn't even spell the name of my school.' On each examination he had to write his name and the name of his school, which was Holmleigh. 'Well, how many ways can you spell "home"?' said John. 'Four. How many ways can you spell "leigh"? Four. So I "permed and conned", hoping I would get one of them right.' Two words, each with four possible spellings, generate quite a number of permutations and combinations.

> 'The headmaster said I was practically illiterate, I couldn't even spell the name of my school'

Writing essays was a nightmare and at the end of every term, when his headmaster reviewed his academic progress, he would threaten John with SNT, short for 'Stick Next Time'. When he got into Cambridge to study natural sciences, 'it was a great surprise to everybody', including him.

'Did you have to work harder than the other kids around you?' Jim asked.

'It's different, yes. You have to find a way to overcome your disadvantages and turn them into advantages. I found that I could think in 3D and I was always amazed that others couldn't.'

> *'You have to find a way to overcome your disadvantages and turn them into advantages. I found that I could think in 3D and I was always amazed that others couldn't'*

At university he made a point of doing things he'd never done before. He had already flown solo in a glider, so took up mountaineering and other adventurous sports. And in his second year, having specialised in geology, he 'managed to talk [his] way on to the best possible adventure, the Cambridge Spitsbergen expedition to Svalbard in the northernmost part of Norway ... It was the last of what I would call heroic exploration,'[1] he said. 'We had no maps of where we were going. We had no photographs. The only thing we had were charts of the coastline

---

[1] A family holiday in North Wales several years later proved far more dangerous than this expedition to the Arctic. John fell down a cliff and was knocked unconscious for three days with a cracked skull and a broken back, necessitating three months in traction and another three months in plaster from his neck down to his thighs.

made by British sailing ships in 1840.' Mobile phones and GPS were a long way off in 1958. They took all their kit 'twenty-five miles across the ice cap with skis and sledges' and transported enough food to keep six people going for three months, while they camped 600 miles from the North Pole and studied rocks.

Convinced that he needed to do more of this kind of thing, he applied to the British Antarctic Survey to do a PhD about the geology of Antarctica. His wisdom teeth and appendix would have to be removed, to eliminate the possibility of either operation being required while he was away. (It was, he thought, a small price to pay.) But the programme was cancelled at the last minute, leaving John desperately disappointed and unemployed. Reluctantly he got a job with his father's company, 'being paid the same as a local teacher'.

Growing up, he had idolised his father, Eric, but their relationship was 'not easy'. John was evacuated to Canada during World War II, aged just three, and didn't see his father for six years. Eric had created the first windproof and waterproof cotton cloth (used in a flying suit for Amelia Earhart, the first woman to fly solo across the Atlantic). During the war, he invented a snap-action thermostat to be used in flying suits for jet pilots. (A suit with enough heat to protect the crew in temperatures as low as minus 60 centigrade would be unbearably hot in the more clement conditions closer to the ground.)

Most simple thermostats work like this: two strips of metal, typically one copper and one steel, are joined together to form what's known as a bi-metal strip. When the bi-metal heats up, the copper strip expands at a faster rate than the adjacent steel strip, forcing the entire thing to bend. This bending action breaks the circuit. It also results in radio waves being emitted when the contacts arc, which would have made the jet fighters visible to the enemy. Eric's snap-action thermostat avoided this problem. It was made from a stressed bi-metal disc that 'snaps open' rather than a bendy strip.

The war ended before the design and new production methods were completed, but hopeful that his ingenious snap-action thermostat could have other applications, Eric set up his own company, Otter Controls. And while his father was busy in his workshop on the Isle of Man, 'trying to find ways to increase and decrease the temperature and the amount of snap', the ten-year-old John was watching him work. 'Why don't you just put a screw in?' John suggested. And his father said: 'Oh, that's a good idea.' 'So if you look at patent no. 600055 and find screw no. 28, that's my idea.'

Despite this early exposure to the joys of making thermostatic controls, John had no desire to work for his father's company when he was in his twenties. 'I was advised that the last thing you should do is go into a family company,' he said. 'Because all the incumbents – directors, managers and heads of department – would immediately think you had come to take their job. And it would make life difficult for you.'

But when the opportunity to spend three years in Antarctica studying for a PhD fell through, he was very short of options. He joined Otter Controls. 'And that's indeed what I found. Being the boss's son made me a bit of a pariah.'

The work was interesting, however. John was asked to make a control that could be used as a time-delay switch for Lightning fighter jets, to stop them accidentally ejecting the pilot when the aircraft was turned upside down and the electrical current surged. John designed a miniaturised control that would do the job. 'Normally you make a temperature switch,' he said. 'And then you have a little heater to make it current sensitive.' If there was a current surge, it would heat the heater. The heater would then heat the bi-metal dish, causing it to snap, and the circuit would be broken. Instead John made the current run through the bi-metal and 'did away with the heater'. If the current increased and the circuit needed to be broken for safety reasons, the bi-metal would be heated up by the surge in the current flowing through it. There was no need to have a separate heater to activate the bi-metal. 'And when I showed it to people they said, "Gee whizz! That's small." So I called that the Otter G control! And fifty years later, the Otter G is still in production!'

Like his father before him, having found the solution to one problem, he started thinking where else his innovation might be useful, and spotted an opportunity to protect motor car windscreen wipers. If windscreen wipers stop moving (when the screen becomes dry or they are blocked by a build-up of snow), it results in a dangerous increase in the current which could overheat the motor, starting an electrical fire. Using his G control, the circuit that powered the windscreen wipers would be cut when the wipers ground to a halt, and so electrical fires could be prevented.

As cars became more sophisticated in the 1960s, electric motors were introduced to do all sorts of different things, not just clean windscreens and start engines. He went on to design a whole family of Otter G controls.

'Now you've got something like twenty-four electric motors in a big motor car with window lifts, seat lifts, and squirter motors for the windscreen. There are motors for washing the headlights and levelling the suspension. You name it, it's got an electric motor in it these days,' he said.

'And all these use this thermostatic safety control that you invented?'

'Or are derived from it, yes.'

The Otter G, invented by John in 1965, is still very much in use today. About 600 million controls have been sold. On average, a quarter of a million units have been manufactured every week for the last fifty years.

'When and why did you turn your attention to kettles?' asked Jim.

Early electric kettles often boiled dry. They did not whistle on the hob and when tea makers were distracted from the task in hand or forgot to fill the kettle with water before they switched it on, the element would burn out. 'It was standard practice that when you burnt out your element, you went into a DIY shop, you got a new one,' John said. 'You took it home, screwed it in, and off you went again.' This seemed unnecessary and wasteful to John, so he invented a thermostatic control that would break the circuit as soon as the element started to overheat and so stop it burning out. His dry-switch-on protector saved consumers a lot of money that had previously been spent on replacement elements. But the big prize, for the kettle manufacturers, was the steam-control switch. 'In the 1960s there was only one automatic electric kettle. All the others just boiled until you switched them off, or until the wallpaper fell off the wall,' John said. State-of-the-art Russell Hobbs electric kettles would switch off automatically when steam was detected in the spout. All other kettles had to be switched off by hand.

'[Russell Hobbs] had the patent and everybody knew you couldn't get around their patent.' Undeterred, John thought he would have a go. 'I never take anybody else's word for it,' he said. 'So I got a copy of the patent, and read it to see what I could do. I've yet to find a patent that I can't find my way around.'

'I've yet to find a patent that I can't find my way around'

'You see it as a challenge, do you, to look at a patent very carefully and see if there's an opening for you?' Jim asked.

'Well, if it's in my field, then yes, I try and find a way around the patent!'

The Russell Hobbs patent specified a mechanical means of showing whether the switch was on or off. Typically, a red button would pop out. 'So I put in a neon light, which was an electrical means of showing whether the switch was on or off,' John said, pleased as punch. He found another loophole too and so was able to work around the Russell Hobbs patent and invent his own steam-control switch which he sold to the other kettle manufacturers. Kettles that switched off automatically sold like hot cakes and soon all electric kettle owners were liberated from the chore of watching kettles that seemed never to boil. And there were other advantages too. Kettles are notoriously power hungry[2] and reducing the amount of time they

---

2 When major sporting events are broadcast on TV, there's a noticeable surge in the demand for power from the National Grid at half-time, when viewers get up to switch their kettles on.

were switched on to an absolute minimum saved a lot of energy.

Electric kettles that switch off automatically are radically more energy-efficient than the old-fashioned whistling-on-the-hob kettles which, even now, remain popular in the USA. Every unnecessary minute that a kettle is left boiling wastes energy. 'I've calculated that if you could do away with all American kettles overnight and replace them with British kettles which switched off when they boiled, you could switch off all the American nuclear power stations and not notice the difference,' John said.

Elements equipped with John's controls sold well but when manufacturers started making plastic kettles, a new problem was introduced. Metal kettles (made of copper or stainless steel) maintain their shape, however hot they get. Overheated plastic kettles melt. If a plastic kettle was left on by mistake, it could melt on to the kitchen work-top or worse, burst into flames. And the British Standards Institute wanted to know: 'What would happen if the dry-switch-on protector failed?'

Kettle manufacturers liked to pass this problem on to the element makers, telling them to make thinner elements that would burn out if the kettle was switched on without any water in the jug. Philips adopted this approach and 'were in terrible trouble with these elements that were burning out,' John said. 'They just weren't reliable. Something like a third of all their kettles with these elements were being returned within the guarantee period. You can't live with that!'

As far as John was concerned, designing elements that

were less resilient was a retrograde step. His dry-switch-on protector had been designed to make elements last as long as possible. It was ridiculous to manufacture elements that were deliberately frail. Instead he proposed a third-level bi-metal control that would move into action if the dry-switch-on protector failed. He combined all three of his kettle safety controls (the dry-switch-on protector, the automatic steam-control switch-off and the new fire-safety control) into a single device. This resulted in a much more efficient design. (John is always looking for ways to get 'every piece to do at least two things'.) When he took the idea to Philips, they 'bought it from the drawing board'.

'Being a very efficient company', Philips wanted the prototype to be delivered as soon as possible and arranged for it to go into production immediately after it arrived. John and the team set to work. Everything was going according to plan. The prototype integrated control was installed in an element and it worked well. The kettle switched off when the water boiled or if there was no water in the jug. Next, mimicking a scenario that could lead to a fire, they disabled the dry-switch-on protector in the prototype, made sure there was no water in the jug and switched the kettle on.

The third-level bi-metal switched the element off, just as it was meant to, and the team breathed a sigh of relief. 'Wow, another successful invention!' they thought. 'We were just about to get everybody to come out to the pub to celebrate,' John said. 'Then we thought perhaps we should let it cool down.' Seven minutes after the invention had been declared a great success, the workshop started to fill with the unmistakable smell of burning plastic.

The new bi-metal control they had designed to stop plastic kettles from catching fire was itself a fire hazard.

The copper element had got very hot (copper being a good conductor of heat). And the heat had then soaked back into the head of the element, which was where John's safety control was located. The heat caused the plastic in the control switch to melt, rendering it useless. The circuit had been successfully broken but, when the plastic melted, the electrical contacts reclosed. 'It was a disaster!' John said.

'We had an emergency meeting and one of the designers said, "We'll have to start again! It doesn't work." To which the gruff Lancastrian managing director replied: "Well, you'd better bloody fix it, then!"'

Philips had pre-ordered 300,000 controls from John and were preparing to go into production. John, meantime, had nothing to offer them.

'I have a theory,' he said, 'that if you take what's gone wrong and turn it into an advantage, then you've got a really good invention.' Rather than focus on trying to stop the plastic from melting, he turned the problem on its head and thought, 'How could we use that?'

'I suddenly thought, "Why don't we put a piece of plastic on to the hottest part of the element so that it would melt first?"' If a strategically placed piece of plastic was attached to a coil spring, then the more the plastic softened, the more the spring would open up. This would force the contacts that connected the element to the power supply to move apart. The circuit would be broken and the element would be switched off.

He didn't think twice about ditching his original idea. He happily abandoned his plan to make another bi-metal control in favour of a wonderfully simple thermal fuse designed around the simplest of ideas: plastic melts when it gets hot. It worked like a dream and had another great advantage for John. Adding a tiny piece of flame-retardant plastic cost a lot less than his original design idea and so reduced the manufacturing costs.

'Within about ten days, we'd got samples. And we tested them and they worked.' John and the team delivered, despite this near disaster. And a million new, improved Philips kettles went into production. Before long, his new company Strix were selling a new integrated safety control to all the leading kettle manufacturers in the world, adapting his design to work in as many different electric kettles as possible. About three quarters of all the electric kettles sold worldwide were fitted with this device.

'How much of your success as an inventor has been due to technical brilliance and how much to being essentially a good businessman, someone who can understand patents and can spot an opportunity?' asked Jim.

'I think it's both. There are a lot of inventors who think that the world will kick a path to their door and beat it down just to have their invention. Life isn't like that. You can't sell ideas, because everybody's against what you're trying to do. People think that other people will want to make their inventions for them. They don't. They haven't got the drive. They haven't got the wish to make it succeed.

'You can't sell ideas, because everybody's against what you're trying to do'

'Nothing is easy in an invention. It's something new, by definition. It's difficult to make and you've got to do it yourself. And you've got to put your money where your mouth is.'

And when you do have a new product you have to be clever about how you sell it. Most people don't get excited by technical drawings, or tiny thermostatic devices, however clever. So John would put his inventions inside a model kettle and present the kettle to the client packed in a luxurious teak box.

'You should never take a new idea to the buyer of a company,' he said. He would explain how his new control might work, but generally found that 'people haven't got the imagination to see how they would use it . . . Instead I would get a design company to make me a beautiful new shape of a kettle and I'd put my new control into the kettle. Then I would take the kettle to the marketing director.' Impressed by an exciting, modern appliance design, 'he would say "Wow!" And I would know we'd sold the idea for the new control system.'

John likes to see an idea through from start to finish. 'You can't separate an idea from its manufacture,' he says. He is deeply sceptical of venture capital. So much of the skill and innovation comes in making things work at the manufacturing stage and, he believes, the people who come up with the idea in the first place are normally the most able to find solutions.

'Britain has historically punched above its weight with inventions. Is this something you think we're losing?' Jim asked.

'I think that the invention industry (for want of a better word) unfortunately has been taken over by the financial industry. I've never borrowed a penny from anybody. You don't need to borrow money to get something into production. If you start small, get an income stream and grow from there, you don't have to borrow money. And you retain all your intellectual property.

'I've worked all my life in companies, manufacturing companies, which have always grown between twenty-five and thirty-five per cent every year. That's usually considered impossible, but you do it . . .

'You don't have to be a rocket scientist or a brain surgeon to move the world forward. There are so many things that are badly designed if you look around.' And even when the designs are perfectly adequate, there are things you can do. He keeps his clients happy by constantly suggesting innovations. A limescale filter in the kettle spout, for example, to stop the scum from getting into your cup of tea. Or a transparent window illuminated with an LED so that we can see how much water is in the kettle. 'If you can make the same product, for the same cost, but with extra features, that will move the market forward.'

'You don't have to be a rocket scientist or a brain surgeon to move the world forward'

John's inventions have helped to transform the design of kettles worldwide. We have John to thank for cordless kettles. 'There was nothing wrong with a kettle that was attached to a plug,' he said. 'There was nothing wrong with a kettle which attached to a base at the back.' But a kettle that can be placed on its base from any angle is more desirable. 'You move the market forward. And you can always do that.'

'If I see a problem . . . I can either think of how to solve a problem in fifteen seconds, or I can't think of doing it at all . . . It helps that I see things in three dimensions. I can make things work in my mind. I can see them operate.

'I observe things as I walk around. I notice things. And

then I am curious about how things are made . . . I don't say to myself, "I must pick that up and look at it", it's just natural. And every time, I think, "How could it be improved?"

'I'm not deeply religious, but I do think you should leave the world a better place than you found it.'

'I'm not deeply religious, but I do think you should leave the world a better place than you found it'

JOHN GURDON

'You always start with a question'

Grew up in: Hampshire
Family: married to Jean with two children, Aubrey and William
Occupation: biological scientist
Job title: chief technical officer at McLaren Applied Technologies
Inspiration: watching caterpillars transform into moths
Passion: trying to clone tadpoles
Mission: to find out if all the cells in an organism contained the same genes
Favourite invention: the nuclear transfer technique that made it possible to clone a frog and paved the way for stem cells
Advice to young scientists: 'Don't give up if it doesn't work at first'
Date of broadcast: 18 December 2012

Professor Sir John Gurdon cloned a frog. Working alone in Oxford in the early 1960s, he proved that it was possible to clone an animal, several decades before the much-celebrated Dolly the sheep. He did this by taking nuclei out of fertilised egg cells found in frogspawn and replacing them with nuclei taken from the cells found in a tadpole's intestines. This groundbreaking experiment in 1962 paved the way for the creation of pluripotent stem cells in the laboratory: cells that multiply indefinitely and have the potential to develop into any type of cell. And it's hoped that therapeutic cloning might one day be used, among other things, to replace cells that have been damaged by heart disease or Parkinson's disease.

Jim Al-Khalili talked to John soon after he had been awarded the 2012 Nobel Prize for Physiology or Medicine, together with Shinya Yamanaka.

In his bedroom at Eton College, John Gurdon spent a lot of time feeding caterpillars in jars and watching larvae turn into butterflies. He was fascinated by insects and plants, but his biology teacher was famously unimpressed. At the end of his first year studying school biology, John (who later won a Nobel Prize for his biological research) was placed at the bottom of his year group of 250 pupils and received an extremely unpromising report. 'He will not listen but will insist on doing things his own way,' his disgruntled teacher wrote. 'I believe he has intentions of becoming a scientist: on his present showing, this is quite ridiculous ... If he can't learn simple biological facts, he would have no chance of doing the work of a specialist, and it would be a sheer waste of time, both on his part and of those who have to teach him.'

'How did you react to [this report], at the time?' Jim asked.

'Well, I wasn't in a position to make a response; it was just presented as a fact.[1] I was informed of the situation, not asked to contest it.'

'So you were asked to give up science?'

'I was *told* I was going to give up science.'

He gave up all science subjects, as instructed, and studied classics at A Level instead. The following year, a private tutor taught him the basics of biology and helped

---

1 The housemaster said, 'Well, one thing at least is clarified: the matter of what you'll specialise in. We can start by crossing out science and then we'll think about what else you might be able to do.'

'We all start life as an egg'

him to get the qualifications he needed to study zoology at Oxford University, not classics. 'I was very lucky that my parents could afford to do this.' His interest in insects persisted and he spent a lot of time as an undergraduate in Wytham Woods, chasing butterflies and collecting larvae, which he would then take back to nurture in his room. When he was not busy being an amateur lepidopterist, he liked to drive very fast cars.

He applied to do a PhD in entomology but was rejected, and he accepted an offer to study developmental biology instead. 'I think the thing to keep in mind,' John said, 'is that you always start with a question. You don't really just probe around and see what happens. We had a very exact question in the 1960s. That question was: Do all cells of the body have the same sets of genes? It is well known now – axiomatic, in fact – that every cell with a nucleus in an individual plant or animal organism contains a complete set of genetic instructions, the same genome. This DNA is what makes you different from me and is present in every one of our cells, be they hair cells, liver cells or fingernails. But this was not known at the time, and seemed unlikely.'

'Why were you so interested in that particular question?' Jim asked.

'Well, that's a very fundamental question in development. We all start life as an egg, and the egg develops, in most cases, without any external instructions – mammals may be a bit different, but for most animals, the egg just turns into a normal individual. And you wonder how that is done.'

When an embryo grows from an egg, cells divide and

multiply. A cluster of generic embryonic cells divides to create daughter cells that are more specialised. As the adult animal starts to emerge, this process of differentiation stops. Skin cells divide to become skin cells. Heart cells make more heart. (Embryonic cells have the potential to become anything. In adult animals, cell identity becomes fixed.) This much had been observed. 'But as late as the 1960s, we had no idea how such a thing could be achieved.' How did the genetic instructions in each cell mastermind this process? How does a frog egg sitting on its own in a pond know how to change into a tadpole?

'Why frogs?' Jim asked.

Frog eggs have always been a firm favourite for people trying to understand how development works,[2] because 'the eggs are big and rather easy to work on'. A frog egg is 4,000 times bigger than mammalian eggs – including those of humans, mice and elephants. 'If you are trying to do experiments where you are moving things around, like transplanting a nucleus, that really makes a big difference,' John said.

John's PhD supervisor, Michail Fischberg, had invited him to repeat a famous experiment that been done by using the nuclei of the snow leopard frog in 1952. Two well-respected scientists, Robert Briggs and Thomas King, had tested a hypothesis that was popular at the time: the idea that specialised cells had lost their ability to divide into anything other than what they were, because the genetic material in the nucleus that enabled them to do this had either been lost or permanently disabled. They did this by transferring the nuclei from cells found in the intestines

---

2 Luigi Spallanzani's experiments in the 1850s involved dressing frogs up in taffeta shorts to see if sperm was needed to fertilise the eggs in frogspawn.

'Actually, it's of no scientific interest to make a clone'

of a tadpole and putting them into fertilised frog eggs which had had their nuclei removed. 'All the business about cloning – actually, it's of no scientific interest to make a clone,' said John. 'What's of interest is to know whether all cells have the same genes, and whether you can reverse the process of differentiation, or specialisation, as it's called.'

After a series of careful experiments, Briggs and King had concluded that tadpoles could not be cloned. When nuclei from very young embryos were transferred, some of these modified embryos grew into healthy tadpoles. But when nuclei were transferred from older embryos, fewer and fewer survived. When nuclei were taken from embryos that were just starting to grow tails, none of the modified embryos survived. These results confirmed the view (endorsed by Briggs and King) that specialised cells lost their ability to become anything else. This seemed, at the time, to provide a perfectly sensible explanation for how development worked.

The task that John was set was to repeat the Briggs and King experiment using a different frog, the South African frog (*Xenopus laevis*). He followed the method that Briggs and King had described, and started trying to extract the nucleus from his frog's eggs using a micropipette. The opening on the end of this intricately manufactured glass dropper had been carefully designed so that it would suck up only one part of the cell, hopefully the nucleus. And the hole it made in the cell wall was absolutely tiny. Try as he might, John couldn't get the micropipette into his eggs. The eggs of his South African frogs were 'covered with a very viscous, sticky jelly' and quite unlike the eggs of the snow leopard frog that had worked so well for Briggs and King. 'They were completely impenetrable by a microneedle,' John said. 'So that was a major problem.' He had to push quite hard to get through the sticky gel, but if he pushed too hard the micropipette went straight through the cell and out the other end, pulling the sticky gel through the inside of the cell and creating a bagel-shaped entity that remained sealed off from the outside world.

After too many failed attempts, he decided to try another approach. He put the cells under UV light, a known technique for eradicating genetic material. It seemed to destroy the nucleus. Amazingly, it also dissolved the sticky gel that coated the egg cell walls. Two problems were solved for the price of one, and courtesy of this happy accident, John was able to insert the new nucleus using a micropipette, just as Briggs and King had done.

'We're talking about removing the central nucleus from one cell and putting it into another cell,' Jim said. 'It's intricate work! Does it help to have a really steady hand and a good eye?'

'Indeed,' John replied. 'Everything you do is done under a microscope, of course. When I was small, I spent an enormous amount of time trying to build a sailing boat in a walnut shell, and the moths I used to collect got smaller and smaller, as time went on, so I think I must have had an inclination to like to do things with my fingers.'

Once he'd overcome the problem of the sticky-gel cell wall, he started to make progress.

'When I was small, I spent an enormous amount of time trying to build a sailing boat in a walnut shell'

'So you were taking the nucleus out of a cell in the intestines of a tadpole and putting it into an egg from another tadpole, to see if it would grow into a healthy tadpole. Or whether, as Briggs and King had shown, that was not possible,' Jim said. 'Were you expecting to get a similar result?'

Both John and Michail had fully expected John to get the same results as Briggs and King. Gradually, however, John found that his results didn't quite match: 1.5 per cent of the embryos that John created by transferring nuclei from intestine cells into older embryos grew into healthy swimming tadpoles. A 1.5 per cent success rate was low. Easy to dismiss, perhaps. But it was a result that directly contradicted the findings of Briggs and King. 'That, of course, makes it much more intriguing. You realise you may be on to something more interesting.'

When John first spotted a normal healthy tadpole, he didn't think too much of it. Given how things had panned out for Briggs and King, it was almost certainly due to experimental error, he thought. He had probably just failed to remove all the original genetic material when he zapped the cells with UV light. Happily for John, while he had been busy trying to penetrate the sticky egg cell walls, Michail had invented a new technique for marking genetic material. Using this technique, John was able to show that none of the original nucleus had been accidentally left behind after the cells had been treated with UV light.

At this point, the fact that he had managed to grow healthy tadpoles became considerably more interesting. Perhaps it *was* possible to clone a frog?

Once he had systematically ruled out all the possible

experimental errors he could think of and was therefore confident that his method was robust, John published his results and reached a bold conclusion. It was possible, if difficult, to clone an adolescent frog. And if this was true, then it seemed highly likely that the specialised cells of his South African tadpoles must contain a complete set of genetic instructions in their nuclei – a possibility that Briggs and King had famously ruled out.

'I can imagine myself thinking, "I must have made a mistake",' said Jim. 'Especially since it was Briggs and King, who are illustrious and established scientists in the field – here you were, contradicting their findings, whilst still a PhD student at the time.'

'That's right. It put one in an extremely precarious position,' said John. 'I had great respect for Briggs and King. Particularly Briggs, who was a wonderful person and I liked him a great deal. He was absolutely correct. He would say, "Well, our results look like this. Yours are different. We'd better see why they're different and which one might be nearer the truth." But there was a time when there was strong criticism, especially from some of the junior members of the Briggs lab, who said, "There must be something wrong," and kept producing various reasons why our results might not be correct.'

'You published a paper on this work in 1962. How long was it before your results were widely accepted?' Jim asked.

'About ten years,' John replied. It was another four years before John showed that the frogs he had cloned were capable of having children of their own. Until he could show that his clones were able to reproduce themselves, both as males and females, he could not conclude that they were fully normal. He couldn't be sure that he had created fully functioning clones.

The theoretical implications of John's experiments on the South African frog were profound. It now seemed highly likely that all the genetic material in the nucleus of these intestinal cells remained intact after these cells had specialised. None of it had been lost or permanently disabled, as Briggs and King and countless others before them had imagined. The answer to the question that prompted John's PhD was almost certainly: 'Yes, all cells do have the same sets of genes.' It's an insight that is so fundamental to modern biology (and was confirmed when, decades later, we were able to read DNA) that it is hard now to imagine a time when this came as a surprise. But in 1966 it was a truly radical suggestion. And it took time for the idea to sink in.

'The most scientifically interesting idea in all of this is that a mature nucleus [taken from a more established intestinal cell] could be rejuvenated when placed in an embryonic cell.' John's experiments showed that scientists could turn back time. (When the nucleus from an adult cell was put in a much younger cellular environment, it started to behave as if it was young again and instruct the embryonic cell to divide and differentiate, rather than just making more intestine cells.) It was a radical insight that took several decades to come into its own.

Forty years later, Shinya Yamanaka (a Japanese doctor who turned to research when medicine failed his dying father) decided to see if he could turn adult skin cells into embryonic stem cells.[3] John had shown that the nucleus from a cell taken from a tadpole's intestine could be reprogrammed when it was placed in fertilised frogspawn. Shinya wondered if he could do something similar with human skin cells and so avoid having to use embryos to create stem cells, which was difficult to do and ethically controversial.

After six years of trying, he succeeded (much sooner than anyone, including him, had expected). And it opened the possibility of therapeutic cloning, 'which is currently attracting a lot of interest', John said. 'This idea ... emerged gradually, and very significantly after the celebrated work of Yamanaka in 2006. [He] really opened up our awareness of the potential for cell replacement.'

Therapeutic cloning involves growing replacement cells in a laboratory. These laboratory-grown stem cells are clones of the patient's own cells and so they are much less likely to be rejected by their host. 'If you can take cells from an individual human and derive other useful cells – whether it's eye, or heart, or so on – then the genetic composition of the new cells and the old cells will be the same, because they are from the same individual originally,' John explained. 'And you do not therefore have to have immunosuppression [which lowers the activity of the body's immune system, leaving it more vulnerable to disease]. That's a major advantage.

'In my view, the most exciting and, in a way, most

---

3 Stem cells have wonderful regenerative properties and open up a range of exciting medical opportunities. They are capable of renewing themselves almost indefinitely (left to proliferate for several months, a small starter population can give rise to millions of cells). And they have the potential to grow into any kind of cell.

imminent kind of therapy in this field has to do with the improvement in vision. There is a disorder called age-related macular degeneration – from which a number of people of my age, more or less, suffer – and the technology which is being tested at the moment provides replacement cells which are simply inserted behind the eye. These [replacement cells] are able to restore some degree of vision, at least in animals.'

'And there are trials going on now in human patients?' Jim asked.

'Yes, and more early next year. I very much look forward to the time when this whole procedure will benefit patients in respect to vision, and then I think it will, I hope, go gradually onward from there.'

Half a century after John published his paper proving that it was possible to clone an adolescent frog, John and Shinya were awarded the 2012 Nobel Prize for Physiology or Medicine.

'What were you doing when you heard about the Nobel Prize?' Jim asked.

'I was in my lab, as I usually am quite early in the morning. The first call that came was from an Italian newspaper saying, "What about a Nobel Prize?" So I said, "I don't know anything about that at all." They seemed surprised and then said they'd call back later. And then, in due course, the Swedish gentleman called. Of course, your first reaction is to wonder if it's a hoax, if it's a joke. You were taught that whatever we get in the lab, you always question it, as you will well know. So you start probing around and thinking, is that likely to be true, or a joke? Some people say, "Well, how do you test that?" Well, it's not really easy to do a test of that.'

'Well, I guess it's reproducibility – you wait for another phone call!' Jim said, smiling.

'Yes, that's essentially right. Most of all, you wait until they put these things up on the web. At one point he said that it would be announced to us later, so you look and see if it's really there.'

Thirty years after John cloned an adolescent frog (and long before he was awarded a Nobel Prize), scientists in Edinburgh cloned Dolly the sheep.[4] 'Does it ever strike you as odd that everyone knows about Dolly, and relatively few people outside your field know about your frogs?' Jim asked.

'I don't think it's too surprising, as a sheep is seen to be much closer to a human than a frog, and so the view would be that if you can do these experiments with sheep – and I mentioned that sheep eggs, like all mammal eggs, are about the same size as [the eggs of] humans – it must be possible, in principle, to do it with humans. So that attracted immense interest. Not that at that time, or even since, there has been a very good reason as to why you should want to do it with humans,' John said. 'But it did raise that possibility.'

'Over the years we've moved from cloning frogs to mice, to other, larger animals – there have been failed attempts by some unscrupulous scientists to clone humans – but are there any deal-breakers, any technical reasons why it couldn't progress further?' Jim asked.

'I can see no reason why human cloning, or nuclear

---

4 Dolly was the first clone of an adult mammal. See the chapter on Ian Wilmut (p. 167) for this story.

> *'I can see no reason why human cloning, or nuclear transfer, should not work just as well in humans as other mammals'*

transfer, should not work just as well [in humans] as other mammals. Indeed, there's some very fine experiments done with monkeys. Nevertheless, there is this high yield of abnormal embryos, and it is going to be a major problem to try and avoid that happening.'

'How long would you say it might take to overcome those sorts of problems?'

'When my first frog experiments were done, an eminent American reporter came down and asked me how long it would be before these things could be done in mammals or humans. I said, "Well, it could be anywhere between ten years and a hundred years. How about fifty years?" Turned out that that wasn't too far off the mark as far as Dolly was concerned ... Maybe the same answer is appropriate.'

'So maybe another fifty years from now ...' Jim suggested.

'Perhaps.'

'Before we'd successfully be able to clone a human being?'

'I think the key thing is "successfully", and by success, I mean with a very high rate of production of normal individuals, ninety or more per cent, perhaps even a hundred per cent. You could very likely clone one human if you did thousands of tests now, but that is a minimal success rate.'

These rather desperate odds didn't stop two scientists based at Seoul National University in South Korea from trying. And in 2004 Woo Suk Hwang and Shin

Yong Moon announced in the online edition of the reputable journal *Science* that they had successfully cloned a human embryo. Their claims turned out to be fake.

'But it does suggest – doesn't it – that there are those out there who are going to want to clone humans?' Jim said.

'I'm not sure that we can do anything about that,' John replied. 'And whether they would really want to clone humans ... I'm not so sure ... The thing they should remember is that when you do these cloning experiments ... you do produce abnormal products, Dolly the sheep being one in two or three hundred. All the others were abnormal. And surely no one would want to produce lots of abnormal humans.'

For many people, cloning a human being is one of the most offensive projects that science could ever engage in. 'Do you think to some extent there is a knee-jerk "yuck" reaction to this?' Jim asked. 'People instinctively feel it's wrong – they use phrases like "playing God", and so on ...'

'I think it's a false concern, and I use the analogy of the thin end of the wedge. People say, "You scientists can do this, but who knows what you'll do next, so we'd better stop you quickly before you do anything else that we might not like." I think that's an unhelpful point of view ... Life does depend on taking certain risks.'

'*Life does depend on taking certain risks*'

'Did you ever imagine that you would be witnessing these kinds of debates when you started your research back in the early sixties?'

'I would have to say I did not. But I think we benefit from what used to be called the HFEA – the Human Fertilisation and Embryology Authority – and they work really responsibly. They ask why you want to do it, what is the question you're asking and why would it be useful. If you have a way of answering that question by procedures that are judged to be ethical, then you get permission. I think that's very good.'

Jim queried John's apparent lack of concern about possible future applications: 'I think that a lot of people who may be happy with the current legislation about working on human embryos – the fourteen-day rule – may still feel just a bit weird or uneasy about where this technology may lead in the future.'

'My comment there would be, try to remember the days of Bob Edwards and *in vitro* fertilisation.' John knew Bob when he was doing that work. When Bob first announced that he could grow babies in test tubes,[5] 'he used to receive the most horrendous mail', John said, 'with people saying, "I hope you die in hell, burning, and suffer as much as possible, because all this is un-natural." However, as soon as he succeeded with Louise Brown, it went to ground, and I think these things do go to ground whenever something turns out to be useful.'

'A lot of people would say that there's a big difference between IVF treatment for couples who want to have a baby, and cloning humans,' Jim pointed out.

'Cloning – the way it's used – means making identical twins. We already have identical twins. We just make more of them by cloning. There's nothing genetically

---

5 The first test-tube baby, Louise Brown, was born in 1978.

different, there's no new genetic combinations – you're
simply copying what nature
has already produced.'

'Certainly there's no genetic
difference between cloning and
identical twins, but this tech-

*'We already have
identical twins. We just
make more of them by
cloning'*

nology would change the whole game from reproduction
that we can't control, to one that is totally within our
control,' Jim argued.

'Only if it worked so well that there were no problems
with it. These abnormalities that I mentioned – if that
was completely eliminated, then cloning could be used to
make more humans. But my guess is that people would
see no objection at that point.'

'Do you think people *will* come around?' Jim asked,
wide-eyed with surprise. 'I can imagine a lot of listeners
thinking, "I'd be horrified with that idea!"[6] And I don't
just mean science fiction, where some evil scientist creates
a whole army of identical soldiers. There is something
deeply disconcerting about that notion.'

'It is true that that would be unnatural, as indeed are
many things. Antibiotics are unnatural. I would take the
view that anything you can do to relieve suffering or improve human health would usually be widely accepted by
the public. That is to say, if cloning actually turned out
to be solving problems and was useful to some people, I
think it would be accepted.'

'So if there were clear medical benefits to cloning

---

6 Current UK legislation makes a clear distinction between therapeutic
cloning for regenerative medicine, which is tightly controlled but legal,
and reproductive cloning, which is illegal.

humans in the future, it's not something you would have an issue with . . .?'

'I think that is something that will be interesting to see, but I would expect that's how most members of the public would react.'

'I'm not too sure about that, if I'm honest,' Jim said, sounding nervous.

'They might not, at the moment,' John said. 'IVF had a very bad press at first and now hardly anyone would say they object to it in principle. It's so obviously beneficial. I think the history is that when something is really useful and does good for humans, people usually support it . . . I can't think of an example where there is a really useful procedure which is objected to by people on purely ethical grounds. Often, when I'm giving talks to the general public, I ask people to vote on the following situation: parents happen to have a child killed, and they can't have more. Some skin is saved from that child, and the mother says, "I'll donate my eggs to make another one." If that would work perfectly, would they vote for it or against it? It's the kind of thing where ethicists would say, "Absolutely not." The average vote on that is sixty per cent in favour. So, forty per cent say no. And the reasons for "no" are usually that the new child would feel as if they're some sort of replacement or would not feel valued in their own right. There are kind of psychological reservations.'

'I think it may be a bit more serious than just a psychological block,' Jim said, unconvinced. 'Because the relationship we have with our children is pivotal to the way our society and our culture runs.'

For professor of ethics and perinatology, John Wyatt, 'the biggest danger [around cloning] is naivety: it's historical naivety, it's sociological naivety, it's philosophical naivety.' 'What this new technology is offering us,' he said, 'is a new way of creating babies in which we control very precisely the DNA of our future children, and we make our future children a product of our own wills and our own choices. In the traditional way of having children, we basically had no control over the way that happened. The genetics of the mother and the father are uniquely mixed into a baby. Of course, there are very strong biological and evolutionary reasons why that is the case.'

John Gurdon defended parents' right to choose how to reproduce themselves. If their child died prematurely and they wanted to try and create a clone of that child from one of their skin cells, then they should be free to do so. 'If the mother and the father, if relevant, want to follow that route, why should you stop them?' he said.

Jim was speechless.

CAROLINE HARGROVE

*'I never thought I would work in
Formula One'*

Grew up in: Montreal
Family: married with two children
Occupation: simulation engineer
Job title: Chief Technical Officer at McLaren Applied
  Technologies
Inspiration: seeing what was possible in Formula One
Passion: creating computer models that mimic racing cars
Mission: pushing the limits of performance
Favourite invention: the world's first simulator of a
  Formula One racing car
Advice to young scientists: 'Follow your passion'
Date of broadcast: 20 November 2018

Caroline Hargrove created the first digital twin of a Formula One racing car, building a simulator machine that mimicked the experience of hurtling round a racetrack at 200 miles per hour. A sporty child who had to wear a back brace for nine years, after she was diagnosed with scoliosis aged nine, she fulfilled her sporting ambitions in another way, working first in Formula One and later with Team GB cyclists, designing technology to help them improve their performance in the run-up to the London 2012 Olympics. She joined McLaren Racing as an academic, hoping to get a bit of industrial experience, and stayed with the company for twenty years.

Jim Al-Khalili talked to Caroline just after she had left McLaren to start a new job with Babylon Health.

It was 'a complete accident' that Caroline ended up working in Formula One. McLaren were looking for people with her mathematical modelling skills. Caroline spotted an advertisement in a magazine and thought, 'Ooh, I can do computer simulations.'

'Nothing to do with an actual car, then?' Jim said, laughing.

No. 'I like Formula One, don't get me wrong,' she told Jim. 'I just never thought I would *work* in the field.'

She was a Commonwealth Scholar and a successful Cambridge academic who applied to McLaren thinking she needed to gain some experience of working in the real world, 'which from an engineering perspective is quite important', she said. 'What I didn't realise was that I'd go there and like it so much, I wouldn't come back.'

}

Growing up in Montreal, home town of the celebrated Formula One drivers Jacques and Gilles Villeneuve, Caroline would bicycle with her brother to watch the races. 'There's a bridge very near the racetrack from which you can get a really good view,' she said. It was fun seeing the cars roaring past, but it was far from being her favourite sport. She was an incredibly active child who loved swimming, tennis and skiing. She dreamt of being an Olympic gymnast. 'At one point, I did gymnastics for four hours a day, five days a week,' she said. But, aged nine, 'all of that pretty much came to a stop.' The doctor diagnosed scoliosis, a chronic sideways curvature of the spine, and Caroline was told that she would have to wear a brace.

For nine years, she was constrained for 23 hours day. Realising just how difficult this would be for Caroline, her parents struck a deal: she was allowed to take the brace off for an hour a day, if she wore it at all other times. And she used her precious hour of freedom to learn figure skating and ballet. 'I would have gone mad otherwise,' she said.

Overnight she went from being the most sporty girl in the class, to being treated as if she was disabled. As the child who wore a back brace, she was treated 'quite harshly' by her classmates. 'Maybe it made me want to prove myself,' she told Jim. She was good at maths at school, but 'not maths in the numbers and being an accountant kind of way'. She liked solving problems and decided to study applied mathematics and engineering at Queen's University in Canada, despite knowing 'very little about engineering'. Fortunately, she enjoyed it. Mathematicians have to prove things. Engineers find solutions. Then, having developed a taste for Europe after an industrial placement in Switzerland in her third year, she applied for and won a Commonwealth Scholarship to do a PhD at the University of Cambridge, making computer models of interacting rigid bodies.

Day one at McLaren was daunting. Inside a very ordinary-looking business estate in Woking, Caroline (who really didn't know what to expect) experienced 'quite a wow effect'. Racing cars were suspended upside down from the ceiling and were hanging off the walls of a very grand entrance hall. She 'loved the buzz' but her excitement rapidly transmuted into fear: 'Oh my goodness – how am I going to be able to live up to what I'm seeing around me?' she thought. Visions of the extraordinarily efficient engineers in pit stops, who were able to get a car back on the track in a matter of seconds, were unsettling. How could she ever be that amazing? 'There was a lot of self-doubt . . . You join and think, "What can I possibly do?"'

Fortunately, the managing director at McLaren had a plan for Caroline, and with a clear problem to solve, her nerves subsided. He had come from BAE Systems, where aeronautical engineers tested the performance of jet-fighter designs in simulators 'all the time'. And he thought it would be a good idea to try and do the same thing with racing cars. Many senior members of the company dis-agreed with his vision, dismissing his ambition as wholly unrealistic. Others might have been demoralised by such a negative attitude, but Caroline found it helpful because it 'helped to take the pressure off'. If no one expected her to succeed, then it would matter less if she failed.

'There were many things I wasn't sure about,' she said. How do you convince a racing driver sat in a model cockpit in Woking that they are hurtling

'There were many things I wasn't sure about'

round a race track at more than 200 miles per hour?

Faced with this rather overwhelming puzzle, she started by making very simple computer models of racing cars, painfully aware of their pitfalls but keen to avoid being paralysed by the complexity of the problem. Then, rec-ognising that she knew absolutely nothing about real-life machines, she took a Formula One driver on a tour of the existing simulators. Together they visited machines that had been built to test road vehicles and aeroplanes to find out what worked well and what didn't.

The test driver would strap himself into a capsule in a model cockpit attached to a mechanical platform that moves objects through six degrees of freedom: three linear movements (lateral, longitudinal and vertical) and three rotations along each of these axes (pitch, roll and yaw). For racing drivers, it's all about the pivot when they go around bends – so getting the lateral movements and yaw rotations was clearly going to be important.

One thing, however, was immediately clear: racing drivers really do drive by the seat of their pants. To experience speed, the test driver needed to feel the vibrations in his buttocks. Recreating these good vibrations was therefore a top priority for Caroline. How precisely did they change in relation to the speed of the car?

For drivers to believe that they were accelerating out of bends, they needed to experience the G-force (the force that acts on an object when it accelerates relative to gravity). Astronauts experience high G acting vertically on their spines, an exaggerated version of the downwards force of gravity to which our bodies are accustomed. Racing drivers experience similar rapid changes in the G-force, and they get it in the neck. 'It's really, really demanding on the body,' Caroline said. When they slam the brakes on as they approach a bend, the G-forces intensify. 'At the end of a race, less experienced drivers can't even hold their heads straight because it's so hard on the head.'

To help recreate some of these G-forces, Caroline's team invented a weighted helmet that can fool the brain into thinking it is being exposed to constant acceleration. We experience acceleration in our inner ear, so there was no need for the drivers' bodies to be moved. Inputs to the head could mimic the sensation of moving rapidly through space, while the drivers stayed seated in the same place. 'I'm not saying that it's perfect,' Caroline admitted, but accelerating drivers through space at Formula One rates would have been impossible.

Most of the drivers learnt to handle the simulator in a couple of hours. Some suffered from motion sickness but they normally got over it. If they did feel a little unwell, Caroline found that a chunk of chocolate and 'a little walk around' normally solved the problem. 'It's amazing how these things just make you forget a little bit that you are feeling queasy,' she said.

To avoid drivers treating the whole thing as an armchair exercise or, worse, a joke, the simulator needed not only to feel like the real thing but to look and sound right too. Without the right sights and sounds, there was no testosterone-charged thrill, and Caroline spent a lot of time getting the sounds and visuals right. It was important to create a high-fidelity system for the things that the drivers would notice. Equally important to the success of the project was avoiding over-engineering those bits that didn't matter too much. As long as the images that were directly in front of the test drivers updated faster than their brains could register the change, the visuals would be good enough.

But drivers, she discovered, are highly attuned to the whining and the roaring of the engine. When they put their foot down on the accelerator, they needed to hear the pitch go up. To achieve this, Caroline linked the pedal pressure in the simulator to a synthetically produced engine noise which changed in very precise ways when the number of revs per minute increased. This noise was then delivered to the driver by surround-sound speakers. But even when these careful adjustments had been made, the test drivers grumbled. Apparently it didn't sound quite right. This led Caroline to consider other noise features (stopping short of recreating the sounds of a cheering crowd). In an open cockpit 'the wind is a big thing', for example. 'We spent so much time on all the little details,' Caroline said. 'It's absolutely important.' Long days spent testing the sound system meant Caroline too became highly attuned to the sequence of noises made by the engine. 'It's like music,' she said. 'For years I would go home and go to bed and just hear the *mneer mneer* of the engine.' Just by listening to the engine, she could visualise what the car was doing and would know where it was on the track.

> 'For years I would go home and go to bed and just hear the *mneer mneer of the engine'*

What Caroline was doing was 'not high-end research'. 'It was more about being an early adopter,' she said. For some things they had to wait for the technology to be created so they could move forward. Their first attempts at creating computer graphics of the track were met with disdain. 'I had so many derogatory comments from drivers,' she said. 'It was non-stop. If they didn't like it, you knew it from day one' and they didn't let it go. In the early days, drivers (including David Coulthard) moaned about the quality of the fast-moving pictures that Caroline had created, comparing them unfavourably to the computer games they enjoyed at home. Caroline and her team were creating digital images that could be projected on to a wall using multiple projectors to try and achieve the resolution needed, and it felt miserable to be criticised in this way. (This was before high-definition graphics and large flat-screen TVs and virtual-reality headsets had been invented.) The rapid pace of innovation in the computer games industry, however, helped her in the end. 'We knew that if we waited another year, graphics cards and projection systems would be much better, so we concentrated on other things.' And, in time, the gaming industry developed the technology they needed. 'Building a simulator is not something that comes from one day to the next.' It took four years.

The simulator was developed to make McLaren cars go faster. Engineers could try out thousands of new design features in 'the sim' and only manufacture those that were the most promising. Since testing virtual models of racing cars is a lot cheaper than building the real thing, this saved a lot of money. And it had other uses too. It enabled drivers and engineers to experiment with different set-ups in virtual reality before the starting gun and tweak their performance ahead of time. This is now an established part of the pre-race routine for many Formula One drivers, but it took a while to get their buy-in.

In the early days of the sim David Coulthard 'didn't really like it', Caroline admitted. 'He certainly preferred his PlayStation!' Mika Häkkinen was 'such a racer' that he couldn't cope with the lack of competition. 'He only wanted to go in [the simulator] if he raced another car': Schumacher, in particular. 'So we had to put a Ferrari in front of him and build a model of a Ferrari just so he would try to overtake it.' When Jenson Button joined McLaren, Lewis Hamilton had to persuade him to give the sim a go.

Caroline's team constructed a virtual model of a new track that had been built in Turkey, for Pablo Montoya. Pablo practised in the sim and got to know exactly what to expect. His competitors, meantime, had to learn on the job. Drivers usually go relatively slowly on the first lap and with each lap, they get faster. 'On his first lap out on the Friday, Montoya was about ten seconds quicker than everyone else.' An impressive margin in a sport where success is sometimes measured in thousandths of a second. When he spoke on the radio and said, 'It's like the simulator!', Caroline was over the moon. Her machine helped Pablo Montoya to win the race. 'I know it's one incident,' Caroline said, 'but I remember being really, really pleased at that stage.'

The simulator had come of age and was being 'very well used' when Caroline gave birth to two other precious things: her children. After 'two maternity leaves back to back', she decided she wanted to work 'maybe a little bit less'. She had hoped to continue in her current role but was told that 'she couldn't possibly lead a team, if she worked part-time', which was disappointing. 'I was told that the world of Formula One was never-ending and you had to put in the extra time,' Caroline said. This was something she had always done (often staying late to work with drivers or to get the job done), and her attitude to her work hadn't changed. 'At the same time,' she said, 'I wanted to spend time with my children and that's something I valued a lot.' Faced with Hobson's choice, she decided to work three days a week and to make sure that she 'at least did things that she could do easily'. 'It wasn't quite the ethos in racing', but having lost out professionally, she didn't want to lose out on the home front too. Working less hard was less fun. Doing a regular job is 'not the same when you're used to pushing a little bit harder,' she said. 'But at least I could do the work with my eyes closed and that kept the stress levels down.'

When her youngest child went into reception at school, she was ready for another challenge. Aware that McLaren had just launched McLaren Applied Technologies, which was designed to show that the company could do 'something other than Formula One', Caroline 'got interested, poked around and asked whether [she] could join'. 'It was me proactively saying that this was something I could do.' She kept in touch with the McLaren Applied Technologies team (it was just two people at that time), and when, 'purely coincidentally', they received a request to build a bobsleigh for the 2010 Winter Olympics, she suggested: 'Well, why don't I do that and we'll see how it goes?' It went well. Caroline started investigating a design of a bobsleigh that made the ride much more comfortable. (With fewer vibrations to distract them, it would be easier for riders to concentrate on strategy and improve their performance.) But sadly, Caroline's plans were never realised. Just as she finished her investigations, the financial crash of 2007–8 resulted in the company losing their sponsorship deal. There was no money to build a new bobsleigh for the Olympics after all.

Not wanting all her hard work to go to waste and 'all that information to be lost', she rang around everyone she could think of who might be interested in what she had done. 'I rang UK Sport and got put in touch with

> '*It was me proactively saying that this was something I could do*'

the head of research and innovation, Scott Drawer, who asked, "Are you guys interested in working with us?" "Yeah. Absolutely!"' Caroline said. Quickly, she then set about trying to convince her bosses at McLaren that it was a good idea. 'I'm not sure it's going to bring us any money, Caroline,' she was told, quite sternly. Caroline agreed. Helping Team GB to prepare for the Olympics wouldn't make any money but it would raise the profile of McLaren Applied Technologies (which was then just the germ of an idea). 'And it would be fun.' 'Interesting projects bring in interesting people,' she argued. She would enjoy the challenge and, with any luck, it would attract some talented engineers to the team.

It worked. Under her leadership, McLaren Applied Technologies grew from the tiny team of three (including Caroline) to 550 people. 'So it's quite different from when it started, yes,' she said. The Olympic sailing, rowing and canoeing teams all wanted help. And the Team GB cycling team, in particular, wanted to be more data focused. The existing technology for monitoring performance could deliver an average speed for a given amount of time: useful information for the Tour de France but not much good for sprinters who wanted to analyse their performance around each bend and seconds after the starting gun has fired.

She worked with coaches and sports scientists to find out what information would be most useful and 'built a little device that sat underneath the seat and collected a lot more data': digital sensors that were capable of generating 200 data points per second. Then she set the team up with the data-analysis tools from Formula One racing. 'They were just flying after that,' she said. 'All these things that you don't think twice about in Formula One' were done for the 2012 British Olympic cycling team. 'We helped them to track their performance at the velodrome in Manchester.' Coaches could run different scenarios in virtual reality and analyse where the cyclists had lost or gained time compared to previous attempts. And the team at McLaren Applied Technologies were heavily invested. 'We all watched the Olympics and were glued to it, as a result of this.' When Victoria Pendleton won a gold medal, everyone felt good. 'It was one small thing amongst many things we did . . . But we were so proud.'

For a young woman who spent her teenage years frustrated and unable to fulfil her desire to excel at sport, things turned out rather well.

MARK MIODOWNIK

*'Every head of department I was
under told me to stop doing this'*

Grew up in: West London
Home life: married with two children
Occupation: materials engineer
Job title: Professor of Materials and Society at University
   College London
Inspiration: being stabbed in the back when he was a boy
Passion: collecting materials
Mission: 'to get everyone making things'
Favourite invention: the makerspace
Advice to young scientists: 'Get your hands dirty'
Date of broadcast: 11 March 2014

Mark Miodownik is a professor of materials and society at University College London. He started his working life in a nuclear weapons laboratory studying the properties of some of the most resilient metal alloys ever made, then moved into academia to escape narrowly defined research goals. He has pioneered collaborations with artists, designers, architects and cultural anthropologists, driven by a desire to make sure engineers make things that are appealing as well as practical. Keen to encourage academics to think with their hands, he created the world's first materials library (home to samples of more than a thousand different materials from aerogel to zinc) and built a large workshop in the heart of UCL, called the makerspace. He would like to see a public workshop, where tools and tactile knowledge can be shared, in the heart of every city.

'It's no accident that the ages of civilisation are named after materials,' said Mark. The Stone Age, Iron Age, Bronze Age . . . '[Making things] is where we came from as a species . . . If you don't get into the Stone Age, you're officially an animal.'

'And when you say "materials", you do mean almost anything: metals, textiles, mobile-phone screens . . .?' asked Jim.

'Well, everything is built of something, and that "stuff" is what we study. That's everything from the windows (so called because they keep the wind out, whereas before they used to let it in), to bricks, to the fabric of your clothes, to the lovely, cushy plastic foam in your sofa . . . all the man-made stuff in the world. My view is that this is an expression of who we are . . . These materials are reflections of our hopes and desires . . . [They are] an expression of where we've come from and who we are, in the same way that a novel or a piece of music is a form of human expression. Imagine shutting yourself off to music. Imagine we all just stopped listening to music. People would be shocked, horrified. But we're shutting ourselves off to making, and it's just as important. This whole thing about the three Rs – reading, writing, arithmetic – is totally wrong! It should be *making*, reading, writing, arithmetic. And there should be a workshop in every

'Imagine we all just stopped listening to music. People would be shocked, horrified. But we're shutting ourselves off to making, and it's just as important'

school and every university that promotes this.'

'You're mad about materials,' Jim suggested to Mark. 'Do you find it disappointing that most of us don't really give them much of a thought? We're happy to make use of the materials in the world, but . . .'

'Are you saying you don't care about them?' said Mark, faux affronted.

'Not as much as you,' responded Jim politely.

'Actually, I doubt that. People do say this to me, but when you get talking to them, you do find that they have a set of materials that they really do care about. The conversations – and those things that squirrel out that in-formation – are usually because you hand them a piece of material. You might hand them a piece of brass, let's say, or you might draw their attention to this strange piece of foam on the windowsill. And it will trigger their mind . . . The hands-on is really important – because actually that is part of how we interact with the world . . . So I imagine that you are mad about some material or other . . .' Mark looked at Jim. 'I'm not sure what it is yet . . . Could it be silk? I've seen your silk shirts . . .'

'Lycra shorts, maybe,' Jim confessed.

'You see!' Mark exclaimed. 'Now we're getting to it.'

Mark was 'turned on to materials' by a dramatic incident on the London Underground that made him painfully aware of the exceptionally sharp properties of steel. Standing on Hammersmith station one day, 'a schoolkid in a blazer with a coat on top', he was minding his own business when a man appeared and demanded that he handed over his money. 'For reasons that are unclear to [him] now', Mark didn't want to hand over the little money he had.

The Underground train arrived and he got on, imagining he had had a lucky escape, but then felt a sharp pain. 'So, there I was,' Mark said. 'The doors had shut behind me, and I was bleeding . . . [Later], in the police station, I saw what he'd stabbed me with, which was a tiny piece of metal – a razor blade. It had gone through all my layers of clothing and my skin and had caused me immense harm. It was a tiny, postage-stamp-sized piece of material, and I couldn't understand how something so small could have such a massive impact.' In the months and years that followed, he 'started to get slightly OCD about steel . . . wanting to know where the stuff had come from and how it could be so sharp and strong.'

'I started to get slightly OCD about steel'

Steel, he learnt, is a metal alloy. Depending on the ratio of its two constituent elements, iron and carbon, and the conditions under which it's made, it can be manufactured to have a range of different properties: super-strong, stainless or razor sharp, for example. Inspired by the range of metallic possibilities, he studied metallurgy at Oxford University and stayed on to do a PhD on oxide dispersion strengthened alloys. This family of metal alloys are capable of withstanding phenomenally high temperatures and tremendous force: useful properties for the manufacturers of turbine blades for jet engines and nuclear weapons.

Mark's first proper job was at Sandia National Laboratories in Albuquerque, New Mexico, the main contractor for the US National Nuclear Security Administration. He spent his days thinking about the atomic structure of super-resilient metal alloys. How could he manipulate the arrangement of atoms in these metal mixes – their microstructure – to create new alloys with the exact properties that were needed? Just how much heat could they withstand before they would crack? The last thing anyone wanted was a nuclear deterrent that might not actually work.

'The great thing about working in a nuclear weapons lab,' Mark said, 'is that they have all the kit that you could ever want to create the most amazing materials.'

'Was it part of your job to design a better nuclear missile?'

'Oh no. I was looking at the intrinsic properties of materials. These nuclear weapons labs have budgets as big as the UK science budget. So they have a lot of cash. And they do fundamental science, because if you don't understand the fundamental science you cannot make sure that a missile in a silo is going to actually explode in thirty years' time . . . If we're going to have nuclear weapons at all – and that's a big debate in itself – but if you are going to have nuclear weapons, you had better know that they work.' Once nuclear weapons are installed and ready to be deployed, 'you can't test them any more.' Manufacturers and politicians alike rely on the laws of physics and chemistry to predict what would happen should anyone decide to press the red button.

'Did it bother you at all, working there?' Jim asked. 'It's not the sort of thing you want to divulge at a dinner party. You know, "Where do you work?" "In a nuclear weapons lab."'

'I was quite happy divulging that,' Mark said. 'I mean, look, we need weapons. I'm not a pacifist. The world is full of people who don't really care about us very much and wouldn't mind if they were to just blow us up. We see that every day. I'm not naïve enough to think you don't need weapons. I hope that we can move towards a society in the future that doesn't have weapons, but we need them now. And we've got nuclear weapons now. Whether we should have gone down that road at all, that's another question. But since we have them, I am quite happy saying we need to keep them in a good state, in a safe state, and understand how they work.'

There was no shortage of resources but 'the difficult thing is that it's all restricted.' Research programmes were narrowly defined and tightly focused on specific goals. And, in time, Mark realised it was 'quite a difficult place to actually express what I wanted to do with my life. Some engineers devote their lives to designing and building one tiny aspect of, let's say, a rocket or a satellite . . .' he said. 'I'm not criticising people who do want to do that, but it wasn't me.'

After a short spell at University College Dublin, he became a lecturer in the engineering department at King's College London and enjoyed having the freedom to study what interested him. There was more to life than metal alloys, wonderful as they were. He started to collaborate with Zoe Laughlin, a student at Central Saint Martins School of Art and Design, and Martin Conreen, a designer from Goldsmiths, University of London, and introduced them to the crackle of tin.

Spending time with Zoe and Martin and hanging out with a more arty crowd than he was used to, Mark temporarily abandoned the scientific method and had fun. Licking metals was one favourite pastime, and this particular bit of fun led to an academic paper describing how the taste of food can vary depending on the metal from which the spoon is made. (As anyone who has tried eating a boiled egg with a silver spoon will know, the metal from which a spoon is made can make a real difference to the taste of the food it transports from plate to mouth.) Mark, Martin and Zoe discovered, among other things, that the taste of cream is enhanced by spoons made from copper or zinc, while spoons made from stainless steel and gold do nothing to improve the taste.

People who use materials every day tend to know best how they appeal to our senses. Typically, artists, not scientists, are the experts on the aesthetic properties of things. 'And you want those experts on your team,' Mark said, even for projects that might seem 'just completely technical'. Building an aeroplane, for example. 'How many people think, "Oh my goodness me, I don't really like the way an aircraft feels inside any more, I don't like the way they sound"?' We need 'a whole load more of these kinds of people involved in the design of an aircraft.'

'You've got to understand the human side of materials as well as their fundamental properties'

'Material science is about taking different bits of knowledge – the physics, the chemistry, the biology, the engineering – and putting it together to make something that people feel is what they need in their lives. Mark believes: 'you've got to understand the human side of materials as well as their fundamental properties.'

'It's taken me twenty years to get there,' Mark said, 'but putting those two things together is what I do now.'

'But it does mean scientists and artists working together right from the start?'

'Not just artists. There are lots of people who study

materials; there's anthropologists, archaeologists ... material culture is actually a huge subject area. History. These people all know something about materials that we don't. They've spent their lives learning about them. They all need to be heard. Designers, architects – these people are experts.'

'What do – or did – your other colleagues in mechanical engineering and material science think of your collaboration with artists?' Jim asked. 'Were you a bit of a maverick?'

'I think so, I think that's mostly what people thought. There's a value system in science which this doesn't fit into very cleanly ... And people worry about that, especially in universities.' One senior academic worried that it would be career suicide if Mark strayed too far from the traditional research agenda of a materials engineer. 'You don't want to end up being a character in the corner who is wheeled out on special occasions,' he was warned.

He was in demand at conferences and would always get invited to speak. 'People wanted to be livened up, I think,' Mark said. They enjoyed being challenged to think differently about what engineers were trying to achieve; but they were less enthusiastic about Mark's initiatives 'in the sober light of day' and very reluctant to stump up any cash for his ideas. For most engineers, hard data on measurable things, like tensile or compressive strength, would always trump more touchy-feely concerns.

Mark, meantime, became increasingly convinced that engineers have to make materials that appeal to our senses. 'You can't ignore the sensual aspects of materials. And where you do, you just end up with things that people don't want.' He would spend his weekends writing grant applications for projects to study the

> 'You can't ignore the sensual aspects of materials. And where you do, you just end up with things that people don't want'

sensory and aesthetic properties of materials and proposed wide-ranging collaborations not just with scientists from different disciplines but with artists, designers and cultural anthropologists. For a long time he made very little progress and worried constantly about the livelihood of the PhD students that he hoped to support: 'their lives, their mortgages, are all dependent on you to keep writing grants.'

'For ten years I really struggled,' he said. 'Every head of department I was under told me to stop doing this. They would refuse to sign research grants in this area for me. I was really discouraged. Most of it, I think, was well-meaning, saying, "This isn't the way I built my career, and I feel you're going to end up unhappy and dissatisfied."' He moved from King's College to University College London, hoping for more freedom. Years spent 'running around for grants . . . [was] really exhausting'. When, eventually, UCL said, 'Look, you've got to stop doing that' and offered to core fund these positions by creating the Institute of Making, he felt 'a huge sense of relief'. 'We've turned a corner now,' he said. 'And I am totally grateful to UCL for doing that.'

Once his position as director of the Institute of Making was secure, he started to contribute to engineering on his own terms. Like most academic scientists, he 'rolled out papers' that were published in peer-reviewed journals. But universities shouldn't just be 'turning research money into paper', he said. Some of his research has ended up as 'a meal, a building, or a piece of clothing'. His team has helped to design materials that heal themselves and explored alternative materials for prosthetic hands. Realising that not everyone wants a piece of flexible, durable pink plastic on the end of their arm, they experimented with prosthetic hands made from leather, wood or more exotic materials. 'The outputs of our Institute of Making are things . . . prototypes of things that will hopefully live in the world and have a long life and be the first of many.' The number of academic papers published and how many times they are cited by fellow academics is not the only way of measuring success.

'Is part of the problem that this kind of tactile, hands-on knowledge can't be shared in the way that knowledge in a book is shared?'

'Yes, I think that is a problem. I think it's intrinsic knowledge. How do you define a great designer, whether it be a clothes designer, or an architect, or someone who creates a new phone? They have this sense of how other people are going to feel about something when they touch it, smell it, have it in their pocket, or wear it, or live in it. And that is deep knowledge. But it's not so easy to put [experiences like these] into words. We're not just thinking, seeing beings. We're touching, feeling, sensory beings. Touching things, holding things, running your hands across them, that immediately triggers a whole set of things.' Aware that knowledge is stored in things as well as books, Mark decided to set up a materials library so that hands-on knowledge of materials could be shared. This collection started with the objects that he had found in 'sheds and grottos' all over the world, and expanded to include 2,000 different materials: acrylic, adamite, aerogel[1] and so on. Aluminium samples include a fabric, two powders (scintillated and plain) and a sheet. Zoe made a set of tuning forks to explore the different sounds made by, for example, acrylic, zinc and walnut wood. There are several different types of artificial grass.

Mark, Zoe and Martin started holding meetings to discuss new projects in the Institute of Making and encouraged staff and students at UCL to join in. And they found the ideas they came up with when they were 'surrounded by materials, floor to ceiling', were completely different to the sets of ideas they had when sitting 'in a room with sticky notes on the wall'. 'Totally different,' Mark said. '[The material library] was set up to allow people to turn their imagination into almost anything,' and designed to encourage chemists, biologists, physicists, artists, designers and architects to get creative by jointly imagining what might be possible. Certain properties of materials have to be experienced to be properly understood. 'It's got this fizz in the air,' Mark said.

'We were one of the first to [create a materials library] and it's actually recognised now as the way to communicate materials science,' Mark said. And they have grown in popularity ever since. 'There's not a country in the world without a materials library.' Keen to create an environment in which the tools and skills needed to manipulate all types of different materials could be shared, they equipped the Institute of Making with laser cutting machines, 3D printers, lathes, ovens, sewing machines, an electronic workshop and pottery wheels, among other equipment.

---

1 The aerogel in the materials library at UCL was, at the time of broadcast, the world's lightest solid: a glass foam made with 99.8 per cent air.

'We love to get people hands-on with materials. But then, once they are hands-on, of course they have all sorts of questions about the science,' Mark said. 'No one person can answer all the questions, so you have to have a community' – people who are happy to interact with the makers and answer whatever questions they can. 'It opens up new opportunities in terms of meeting people and the social side, but it also advances your theoretical knowledge. It makes you understand why you need theory. At the same time, it's hugely enjoyable. People love it.'

Like the library, it's open to the staff and students at UCL and Mark hopes to get academics from all sorts of different disciplines thinking with their hands. They also run public events, offering a communal garden shed for city dwellers who typically live in flats, as part of the wider makerspace movement to bring tools and making back into people's lives and create communities of makers where tactile knowledge can be shared.

'There are now about one thousand three hundred [maker-spaces] worldwide,' Mark said. 'I'm really happy about that. And I really look forward to the moment where there is a public workshop in every village, town and city of this country. It's not that I don't like libraries. But I think a public workshop is more important these days than a public library. We can access books from the comfort of our own home on smartphones, but fewer and fewer people, especially in big cities, have access to tools and workshops . . . I would really like to see local authorities out there convert their libraries into workshops.'

'I think a public workshop is more important these days than a public library'

'Now you can understand that that sort of statement could sound quite inflammatory,' Jim interjected. 'You know, people up and down the country are working hard to save their public libraries and keep them open.'

'Reading books, I have no problem with that,' Mark replied. 'But I think usage of that space and that resource could be better. People would come to these workshops and would gain a huge amount from them . . . They would talk to each other and do stuff together, instead of sitting quietly in the corner.

'Making . . . should be part of everything we do. What's at the heart of a home? The kitchen. What do you do in a kitchen? You make food out of materials. What's the heart of a city? I think it's a big workshop.'

'It would be a health and safety nightmare, of course,' Jim said.

'That's what everyone says to us. And the truth is that where there's a will, there's a way.'

'I do see what you mean,' Jim agreed. 'In this digital age, it's not access to information that most people are lacking. It's the space to play around with the actual physical stuff.'

'And [being part of a] community of makers. There are lots of people in their sixties, seventies or eighties who have a huge amount of knowledge about how to make stuff, which they can pass on to the next generation, if only you gave them a place to do it.' Public workshops, or hackspaces, offer much more than a place to fix the toaster. They are places where people with skills and experience can share their hands-on knowledge, make friends with younger people and 'encourage them to appreciate that [making] is another way of understanding the world, and one just as important.'

'So you're starting to *think* in a different way?'

'What worries me sometimes,' Mark said, is that the more articulate members of society 'think that everyone else should just think and speak or read and write.' The people who express themselves by talking tend to end up in positions of power. Meantime, 'the makers often just like being in the lab, making stuff.' Because they don't engage with the debate, their views are marginalised 'and they don't end up making the big decisions.' The space available for workshops shrinks as a result. 'And making shrinks.'

There should be more manufacturing, and less pontificating about the world, Mark feels. Reading, writing and arithmetic will only ever get you so far. 'Making things is who we are.'

'Making things is who we are'

CLARE GREY

'I suddenly realised I didn't need to know the answers'

Grew up in: Middlesbrough, the Netherlands, Belgium and St Albans
Home life: married to bio-physicist Daniel Raleigh
Occupation: materials chemist
Job title: Professor of Chemistry at Cambridge University
Inspiration: realising that research was about asking the right questions
Passion: tracking the movement of atoms inside working batteries
Mission: to find new ways of looking at the chemical processes and use them to help power the twenty-first century
Favourite invention: a new method for studying what happens inside an operating lithium ion battery
Advice to young scientists: 'Start doing research as soon as possible'
Date of broadcast: 6 March 2018

Professor Clare Grey studies what happens to atoms during chemical reactions and has transformed the performance of rechargeable lithium ion batteries. She specialised early in nuclear magnetic resonance spectroscopy, a technique that tells you about the structure of atoms within molecules, and had a paper published in the prestigious scientific journal *Nature* when she was just twenty-two. A chance conversation with a scientist from Duracell sparked a lifelong interest in battery technology. Her research has contributed to some dramatic improvements in the design of the rechargeable lithium ion batteries that have powered the portable electronics revolution. In 2015, Clare's research group developed a new type of 'lithium air' battery. If a commercial application of this battery were to succeed, it would enable electric vehicles to travel much further without needing to be charged.

When Sony invented the first digital video camera (known as a camcorder), they needed a battery that could store a lot of charge and yet be light enough to rest comfortably on a movie-maker's shoulder. 'The lithium ion battery they put together [in the early 1990s] caused a global explosion,' Clare told Jim. 'It formed the basis of the portable electronics revolution.' A battery that was designed for a camcorder is now used to power mobile phones, tablets and laptops. It changed the way we communicate and consume information, and gave us digital cameras, games consoles and electronic toys, hedge trimmers and cordless drills.

'These rechargeable lithium ion batteries really are the unsung heroes of portable technology,' said Jim. 'We do take them for granted, don't we?'

'Yes, absolutely,' Clare agreed. 'And I think there's always an expectation that the chemists are going to do a lot better, and make better batteries as more and more power-hungry devices come along.' If only the people who make batteries could keep pace with the digital revolution, then we could all have the smartphone of our dreams: devices that are capable of lasting for weeks, not days, without needing to be recharged.

'Does it bother you when people start complaining about having to charge their phones more and more often, when in fact battery technology *has* improved?'

'The price of battery technology has gone down by a factor of ten. Energy density – that's how long your battery will last for – has improved by about three times since [the early 1990s]. So, it bothers me. But I think what bothers me more, is that people think it's a simple problem to solve.'

It's not. If a phone battery is going to be charged every day and is expected to last at least two years, 'or maybe seven', then it needs to be capable of charging and discharging at least 730 times. Every time a battery charges, it converts electrical energy to chemical energy. This chemical energy is stored in the bonds between atoms that are created when new molecules are formed. When the battery discharges, these chemical bonds are broken and electrical energy is released. 'And you've got to get those reactions to go backwards and forwards many, many times,' Clare said. (The first rechargeable battery was called the rocking-chair battery because it had achieved this.) 'Very few processes can occur reversibly without failing, or degrading, as we call it in the field. And so it is a real challenge. I think people do need to appreciate that, and maybe stop moaning, when, every now and then, their mobile phone batteries don't work.'

> 'It is a real challenge. I think people do need to appreciate that, and maybe stop moaning, when, every now and then, their mobile phone batteries don't work'

'So it's not how long the charge lasts, it's the battery life [that is the focus of your research]?' Jim said.

'We're pretty good now at holding charge. You don't worry about that. Unless you insist on putting your battery in a freezer or an oven, you can be pretty confident that once you charge it, it won't discharge. It's the small degradation reactions that happen every time you charge and discharge [your phone] that ultimately kill the battery. Those are the sorts of things that I'm interested in. Understanding what's going on, and working out how to stop some of those things.'

}

Clare won an Open Scholarship to one of the most academically prestigious Oxford University colleges, Christ Church.

'You might have been a medic, is that right?' said Jim.

'I was very indecisive during my last year at school. And it was only at the eleventh hour, as my father will remind me, that I finally decided I was going to do chemistry.'

'Once you got to university, did you say, "Yep, this was the right choice"?'

'I had moments where I didn't think it was the right choice. I was in this environment that was very competitive.' As the second ever woman to study chemistry at Christ Church, she felt very exposed. 'It was an environment where you had to know the answers to everything,' Clare said. When she started working on a research project in her fourth year, however, 'the dynamics changed'. Students from all over the world were working alongside one another in the lab. 'It was a nice, supportive group. The activity really was about how the group could succeed, and how to work together to do research.

'I suddenly had this moment where I realised I didn't need to know the answers.' Staring out at the traffic on South Parks Road, it occurred to her that not knowing the answer was 'what research was about.' The most important challenge was to ask the right questions. 'So instead of having to constantly defend myself and come up with answers – often on the fly in these aggressive interactions – it was up to me to define the questions and then think

'It was up to me to define the questions and then think about how I was going to answer them'

about how I was going to answer them.' This was the moment Clare knew that she 'wanted to be in chemistry'.

She specialised in NMR spectroscopy (a technique which exploits the nuclear magnetic resonance of individual atoms) and was using it to study the structure of tin materials. This felt like a useful thing to do because tin materials often make good catalysts.

'And am I right in thinking that, when you were an undergraduate, NMR wasn't used on solid materials?' asked Jim.

'What I found exciting at the time was that people had done a lot of work on solution NMR, and that was emerging as a field.' Solution NMR is the science that underpins MRI scans. (It enables us to create detailed images of the insides of our bodies and to access areas that X-rays and ultrasound can't reach.) Indeed, magnetic resonance imaging (MRI) used to be called nuclear magnetic resonance imaging, but the word 'nuclear' was dropped to avoid the negative associations of the word. When Clare was a student, it was just starting to be used in solid-state chemistry to look at the structures of materials. 'When I moved into the space, the field was completely wide open,' Clare said. 'I think that's what very much excited me and attracted me into the area.'

NMR spectroscopy works like this. The object of interest (be it a body or a piece of tin) is put inside a strong magnetic field, many times stronger than a fridge magnet. This causes the magnetic moments associated with some nuclei at the centres of atoms, or ions, to align with the magnetic field, in the same way that a magnet inside a compass aligns with the earth's magnetic field. The nuclei are then perturbed with radio waves so that they now rotate about the magnetic field – a bit like a gyroscope – giving off radio signals with different frequencies. These are then picked up, and the numbers of ions and atoms in different local environments (which give rise to different signals) are plotted on a graph, known as an NMR spectrum.

'People in the physics community were exploring what sort of things [NMR] could do in principle.' They worked on idealised models of solids. Clare applied NMR to materials that existed in the real world that had atomic structures that were messy and difficult to study. 'Of course, there were other people doing similar things,' she said. 'But we were some of the first to do it in these complex, relevant systems. And then I also showed that [NMR] had potential applications in interesting materials' – from tin catalysts to the phosphor materials that coat the inside of fluorescent light bulbs, for example.

'I should say at this point,' Jim interjected, 'that this work delivered some outstanding results. You ended up with a paper in *Nature*, as an undergraduate!'

'I was very lucky as an undergraduate,' Clare said. 'Luck plays a part in all discoveries, as you know, Jim.' Nonetheless, it was pioneering work. Not many 22-year-olds get papers published in prestigious scientific journals.

She went on to do a PhD in the same field, relishing the opportunity to 'think more deeply about science'. This might have been possible when she was an undergraduate, Clare acknowledged. But she

'Luck plays a part in all discoveries'

was doing other things: rowing, singing, playing the cello and 'generally becoming an adult'.

A Royal Society Research Fellowship gave Clare the freedom to study whatever she wanted, wherever she wanted, and she decided to go to the University of Nijmegen in the Netherlands 'to improve her knowledge of NMR'. In particular, she wanted to see if she could use NMR to work on the distances between ions in solids. She had lived in the Netherlands as a child and enjoyed relearning a bit of Dutch as well as learning more about NMR spectroscopy.

A year later she moved to the USA to work as a visiting scientist for the multinational chemical company DuPont, to find out more about how materials were made. 'I've always been interested in doing something where there is a connection between what I'm doing in the lab and real-world problems.'

> 'I've always been interested in doing something where there is a connection between what I'm doing in the lab and real-world problems'

'What was it like, working with scientists in industry, who were not just purely curiosity driven?' Jim asked.

'Well, it was an interesting time. It was very much the end of the big industrial labs, and DuPont was transitioning from a premier chemistry lab to an industrial research lab, being much more relevant to the business units. I suffered a little bit from that because the projects were changing all the time, which I had no control over. But I also got very interested in some of the environmental issues that DuPont was dealing with at the time.'

A ban on CFCs had been introduced by the Montreal Protocol in 1987. (Scientists studying the earth's atmosphere had discovered that CFCs were speeding up the destruction of the ozone layer, and politicians had taken decisive action.) These liquids and gases are wonderfully inert and had been widely used as industrial propellants (in aerosols for spray deodorants, for example) and as refrigerants. They had provided a safe way of cooling the contents of our fridges for more than half a century. (In the very early days of fridge manufacture, flammable refrigerants caused fridges to randomly explode in kitchens.) But following the Montreal Protocol, fridges were rolling off the production line without a suitable refrigerant to take the heat away. The research team at DuPont came up with an alternative: hydrofluorocarbons or HFCs (created by replacing the chlorines in CFCs with hydrogen atoms).

Clare studied the structure of different HFCs (using NMR to look at the fluorine ions) and used this information to help DuPont find ways of separating mixtures of HFCs and hydrochlorofluorocarbons (HCFCs) produced during the more complex syntheses. Once DuPont had found a way to manufacture HFCs, the next generation of ozone-hole-friendly fridges were rolled out. HFCs had the cooling properties of CFCs without destroying the ozone layer. But sadly, several years later it became clear that HFCs were not as good for the environment as everyone had imagined. They were later discovered to be a potent greenhouse gas, with a global-warming potential up to 3,000 times greater than that of carbon dioxide. And DuPont had to innovate again.

'Many of these companies were complaining about having to change because of an environmental regulation,' Clare said, 'but then they went ahead and made money on it.' New regulations can often be seen as 'stifling industry', but they can also be good for business.

The work at DuPont was interesting and relevant, but Clare didn't particularly enjoy being told to stop working on one thing in order to pursue a different line of research, just as the research she had started was becoming interesting. 'One of the reasons that I didn't want to stay within the industrial environment was that I wanted to be able to define my own research problems,' she said. 'I didn't want to be beholden to the changes or the other motivations that inherently were associated with being in a company.'

After two years at DuPont studying the fluorine ions in HFCs, she got a job as an assistant professor at Stony Brook, the State University of New York. Being an academic in America appealed because you're 'given the kit to get going'. Clare wrote a grant proposal and was given an NMR spectrometer. 'They gave me some money to set up a lab and then I was able to test out some of my ideas.'

'And it was around this time that you met the person from Duracell who sparked your interest in batteries?'

'Yes. It was quite a funny story, actually,' Clare said. She had given a lecture at a prestigious Gordon Conference in New Hampshire, famous for bringing together pioneers from different areas of science. Afterwards a scientist from Duracell told her how much he had enjoyed her talk on lithium ions. Clare had made no mention of lithium in her talk. 'I wasn't looking at the battery space at all,' she said. (She was looking at how fluorine ions move through solids, building on the research she had started at DuPont.) 'But I didn't have the heart to correct him. "Thank you very much," I said. And he asked if I would look at some of his new battery materials. So, I took some of his samples back to the lab and asked my PhD student to examine them using NMR. We realised two things very early on. One: we didn't actually understand what the [NMR] spectrometer was telling us. And two: what was published in the literature didn't have the full story by any stretch of the imagination!'

Interpreting the graphs produced by the NMR spectrometer was more complicated than anyone had imagined for reasons that, at that time, were far from clear. But there was a silver lining. 'Although subtly different', the materials that Duracell had developed belonged to the same class of materials as the catalyst material that she had worked on as an undergraduate and as a PhD student. They were both paramagnetic. 'So I had the right training. But we still had to delve down into what was specific about these systems.' And there was another silver lining. 'It was Long Island, it was winter and at that time I played a lot of squash,' Clare said. Her squash-playing friends worked at the Brookhaven National Laboratory, which was a 30-minute drive away. And, as Clare discovered during a post-match chat, they were working on batteries. Thanks to them, she was able to visit the lab and learn how batteries were made.

'It pays to take time out to play squash every now and again,' Jim said, smiling.

'The conversations after a squash match got me going,' Clare said. 'I don't play squash very well, but still, in this case, it was very productive.'

It took Clare and her PhD student two years to work out what was going on. (Interpreting NMR spectra requires an intimate understanding of the precise structure of materials and how they are likely to behave, and some deep theoretical thinking.) When she felt her results were ready to be published, she phoned her contact at Duracell to tell him the good news. 'We know how to interpret the spectra,' she said, excited. 'Tough, Clare. We've cancelled the programme,' he replied.

'After two years of research, you'd cracked the problem, and they weren't interested!' Jim said, slightly horrified.

'Well, I'd cracked the problem at one level,' Clare said. 'And the fault also lies with me for not necessarily keeping him abreast at every stage of the way, like a lot of scientists do. But it wasn't such a bad story in the end, because he then asked me if I would continue to work with him on primary batteries, the batteries that you use once and then throw away. And he funded another PhD student.'

'It's interesting that Duracell, having been interested in lithium ion batteries, then go back to the traditional battery technology . . .' Jim said.

When, in the early 1990s, Sony began to make real progress with the lithium ion battery, the traditional battery companies – Duracell, Energizer and Rayovac – 'got out'. And focused instead on selling as many disposable batteries as possible.

So Clare studied disposable batteries and, even though Duracell were no longer so interested in rechargeable technology, she continued her work on rechargeable lithium batteries as well. Both involved making a lot of batteries, charging them up 'to different stages' and then breaking them apart to study the materials inside and how they had changed.

'How novel was the research you were doing, using solid-state NMR to investigate what was going on inside batteries?' Jim asked.

'When I started to do it there had been a few studies by other groups worldwide . . . but we were one of the major groups in developing the whole field in a much more systematic way. We were instrumental in . . . really fleshing out the whole of NMR spectroscopy of batteries.'

When Clare started this work, NMR was on the peripheries of battery research. Materials were more commonly studied with techniques such as X-ray diffraction, not NMR. '[X-ray diffraction] gives you an average picture,' she explained. 'What I could do, though, was take the average picture and then work out the fine details. And it's often the fine details that dictate whether the lithium ions move fast in the solid, whether you can pull them out and whether you can charge them.' The more freely the lithium ions can move during the chemical reactions that enable it to charge and recharge, the better the battery performs. 'So we could say whether the lithium was near manganese in its oxidised form or its reduced form. We could say whether the lithium was surrounded by four oxygens or six oxygens. And I know that sounds slightly esoteric, but if you want to try and understand how this very complicated and complex material functions, [it's very useful].'

'I'm trying to picture you in the lab . . .' said Jim. 'How did you actually go about studying batteries?'

'When we started twenty years ago, we made our battery materials or got them from other collaborators.' Promising materials were baked in ovens set to over 800 degrees Celsius and mixed with carbon, 'because the carbon allows you to get the electrons in'. The contents were then clamped together in a coin cell, like 'the little button cells you would put in the back of your camera'. These batteries were then connected up to an electrical circuit and left to charge and discharge, sometimes up to a thousand times. Different batteries would be disconnected at different stages in the charging process and pulled apart to do the NMR. The results of the NMR could then be used to work out what chemical reactions had or had not occurred inside the battery at a particular moment in time.

'So you're carrying out post mortems on batteries? Investigating the cause of death of a battery . . .'

'Yes, but we're not always looking at death.'

'But you're having to open it up, like an autopsy . . .' Jim said, enjoying the analogy.

'Yes, we're ripping it apart. And then, on a good day, we can then reassemble the battery and, unlike an autopsy, the battery continues to function,' Clare said, smiling.

'We're trying to prolong their life but also to optimise them, to make them work better. And there are different definitions of "work better". It could be, you want batteries that charge faster because we'll want electric vehicles that charge faster when you go into the equivalent of the petrol station. But sometimes we're trying to make them last longer.'

'And [by that you mean] last longer over a lifetime, not just hold their charge for longer.'

'It's lasting for longer as you continually charge and discharge, yes.'

Clare explained, 'Some of these battery materials are a bit like sandwich cakes, with sponge and a [jam] filling. And we're constantly pulling the filling out and shoving it back in again. We're doing this thousands of times, and the materials are continually expanding and contracting. And we've got to get them to last for thousands of cycles. That's the challenge.'

'Some of these battery materials are a bit like sandwich cakes'

'Is it all about the jam or the sponge?' Jim asked.

'The jam is the lithium ions, so they're pretty constant. It's the sponge that's the problem. [Problems arise] when the sponge bits get into the jam, and it stops the jam moving around.'

When a battery is first made, the lithium ion 'jam' forms a discrete 'filling'. As it discharges, the lithium ions move through the 'sponge', enabling a current to flow, and the net result is a lot of jammy sponge. In disposable batteries this jam-soaked sponge is thrown away. Or, hopefully, recycled. But if the same battery is to be used more than once, the battery 'cake' needs to have its nice, neat, layered structure restored. 'I make a material and I put it in my battery in the way described. And then I pull the lithium out. And what I don't want is the structure to collapse, I want it to stay the same,' Clare explained. 'And what happens over time is, these structures change. They might go from a nice layered structure, to use the cake analogy, to a messy structure where there's lots of crumbly bits.'

Making sure that all the lithium ion 'jam' is put back in the middle of the 'cake' is a challenge. (Once a cake has been sitting for a while, getting jam out of the sponge isn't easy, and the same is true for lithium ions in batteries.) Typically, about 50 per cent of the lithium ions are left behind after a battery discharges, reducing its capacity to store charge. To make rechargeable batteries that last for ever, charging and discharging *ad infinitum*, the movement of the ions, and any associated change in the structures of the surrounding materials, need to be completely reversible.

Clare developed NMR spectroscopy of batteries, and pushed it as a way to characterise the materials that were used to make them. However, even when she had worked out a way to do this, her job was far from complete. NMR spectroscopy is arcane and difficult to understand, even for fellow scientists working in related fields. It is fiendishly difficult to interpret what the results might tell us about the atomic structure of the material being studied.

'And making that accessible to industry is important,' said Jim.

'If you want your understandings to be understood by the material scientists and the chemists and the physicists, you have to put it into a language that bridges different communities. It's no good going into an electrochemistry conference, and talking about the physics behind the technique, if you can't actually tell people how their material is functioning.'

Clare not only developed this new technique, she created a new language, a system that could translate peaks on a graph generated by the NMR spectrometer into information about the structure of the materials that everyone working in the field could understand. It helped fellow scientists to understand how complex materials function, and it helped companies to test different materials and find those that were most likely to perform the best.

'We helped the Duracell scientists to work on a new manganese oxide which then went into a commercial AA and AAA battery,' she said. 'But I think the most important thing is the idea of the fundamental science. My work has contributed to the understanding [of the atomic structure and behaviour of] of battery materials, which has helped scientists globally to come up with the next generation of [battery] materials.' If we want the 'little rechargeable batteries in our mobile phone that we're irritated with' to improve, one shouldn't underestimate the power and necessity of fundamental science. It's only by understanding, in great depth

'I think the most important thing is the idea of the fundamental science'

and detail, how battery materials work that we are able to move technology on. 'So, I think that's my biggest contribution.'

On a sabbatical in Amiens in northern France in 2007, Clare made batteries in plastic bags. Instead of clamping battery materials in a metal case, the contents were vacuum-packed using exactly the same plastic bags that food manufacturers use to seal fish, smelly cheese or high-end coffee. Industry was interested in this 'coffee-bag battery' technology because plastic bags were lighter and cheaper to make than metal cases. The big advantage for Clare was that they were transparent. Radio waves could pass through the plastic bags, and this meant that Clare could perform NMR on the batteries without having to pull them apart to examine the contents. Excited by this possibility of doing NMR on batteries without having to destroy them, she made her own plastic-bag batteries, ordering the bags online from food manufacturers and sealing them with a laminator.

At the end of her sabbatical, she packed her suitcase full of plastic-bag batteries, flew back to the USA and started doing *in situ* NMR to monitor the chemical reactions and changing atomic structures inside these plastic bag batteries. There was no longer any need to make lots of

batteries and stop them at different points during the cycling, which had been 'a tedious exercise in principle'. 'We
literally put leads down the bore or hole in our magnet
and connected them to the batteries,' Clare said. The
radio communications used by the local ambulance service caused a few problems, and they needed to 'put a bit
of aluminium foil down the magnet' to stop it interfering
with the experiments. But otherwise it was wonderfully
straightforward. A single battery could be charged and
recharged, and radio signals could be collected across its
lifetime providing a record of all the chemical reactions
that had taken place. Battery research projects could be
completed in a fraction of the time. And there was another, unexpected bonus. They found evidence of many
intermediate changes in the chemistry of the battery that
had not been detected before.

The twenty-fifth anniversary of the lithium ion battery was
celebrated in 2016. Jim asked what comes next: 'I know
you've been looking to the future . . . looking beyond lithium ion technology [that powered the portable electronics
revolution].'

'I think we, and others, realise that there are ways we
can improve the current technology. At the same time, I
think it's also very important to look beyond lithium ion,
and look at other ways of storing charge. So my group is
very interested in sodium ion batteries. They are cheaper
because there's a lot more sodium in the world, and they
are quite similar to lithium. So, in many ways it's a drop-
in technology. We're also interested in magnesium, and
again there is more magnesium around. We're also interested in the so-called "beyond lithium ion technologies".

Lithium air and lithium sulphur, but mostly lithium air.'

'In 2015 you built a prototype of a lithium air battery, which caused a lot of excitement,' Jim said. 'I remember reading about it in the news. What is the excitement about lithium air, distinct from lithium ion?'

'It still involves lithium ions, but the distinction is, you're reacting lithium with oxygen to form a lithium–oxygen product.' It works by 'breathing in' oxygen from the surrounding air, and these batteries weigh considerably less than their lithium ion predecessors. When the battery is in use, the oxygen in the air reacts with the lithium in the battery to form a solid oxide – lithium peroxide. When it's being charged, the oxide is broken down, releasing oxygen back into the air. 'Because air is abundant, we don't need to worry about the air. The lithium is light. And the peroxide is light.'

Super-lightweight lithium air batteries could enable electric vehicles to travel further without needing to be charged because cars with lighter batteries are more fuel efficient. There is no guarantee that a battery that works in the laboratory will be commercially viable. And it will take time: 'Twelve or fifteen years, because it's a complex problem involving lots of different things ... But if you can get it to work, you can have the same energy density as gasoline or petrol,' Clare said. 'So, it's the ultimate battery.'

It can be hard for people who are not chemists to get excited about atomic structures and methods such as NMR. 'It may not be transparent to everyone,' Clare said. 'But at least to me, I can see the relevance of understanding how materials change as you [charge and discharge them].' The work done by Clare and her team has transformed the performance of the rechargeable lithium ion batteries that can be found inside every mobile phone. 'The next generation is moving into electric vehicles.'

In future, we will need to find ways to store the energy that is generated from renewable sources so that we can use that electricity when there is no wind, or the Sun doesn't shine. ('People think of the [National Grid] as a way of storing electricity, but it's just a way of pumping it around.') And here, 'batteries will play one part in the solution,' Clare said. 'I think we will also need to think, as scientists, of other ways to convert chemical to electrical energy.' New ways of storing charge are urgently needed. Clare is a founder member of the Faraday Institution, a virtual research hub that was set up in 2017 to encourage scientists and engineers to embrace 'the big battery challenge'. Let's hope they succeed in creating the technologies we need to power the twenty-first century.

IAN WILMUT

*'Dolly the sheep was the first clone
of an adult animal'*

Grew up in: Warwickshire
Home life: married to Vivienne with three children
Occupation: embryologist
Job title: Chair of the Scottish Centre for Regenerative Medicine
Inspiration: a summer job at the Animal Research Laboratory in Cambridge
Passion: experiments to improve animal breeding
Mission: to clone an adult sheep
Favourite invention: Dolly the sheep
Advice to young scientists: 'You don't need to be an academic high-flyer to do scientific research'
Date of broadcast: 11 October 2016

Professor Sir Ian Wilmut created Dolly the sheep. He wasn't particularly academic at school or university, but decided to do a PhD on the preservation of boar semen, inspired by a summer job at the Animal Research Laboratory in Cambridge. He was a junior member of the team that made Frostie, the first calf to be born from a frozen embryo, in 1974. Working at the Roslin Institute in Scotland in the 1980s, he was told to look into the possibility of making genetically modified animals. (Cloning was an important part of this process.) Reluctantly he agreed. After hundreds and hundreds of failed attempts, Ian and his team created the first clone of an adult mammal. They published a paper in 1997 describing what they had achieved, and Dolly (who was then seven months old) became world-renowned.

Jim Al-Khalili interviewed Ian before an audience at the Edinburgh Festival in August 2016 to celebrate the twentieth anniversary of Dolly's birth.

The birth of the most famous sheep in the world was a low-key event. The man who created Dolly the sheep was digging his vegetable patch and other members of the team were elsewhere. Seven months later, in February 1997, Ian Wilmut and his team published a paper announcing the arrival of the first clone of an adult mammal, and Dolly's barn at the Roslin Institute in the lowlands of Scotland became a hive of media activity. Journalists quizzed Ian and the team at the institute about what they had done, and learnt how to photograph a sheep.

'How did Dolly react to all this media attention?' Jim asked.

'Well, she became spoiled,' Ian said. 'The photographers soon discovered that sheep were even less obedient than the people, and that if they wanted her to go to a particular place they'd have to offer her food.' Since Dolly was being photographed several times a day, over a period of months, 'she became overweight'. Looking slightly heavier than usual, Dolly appeared on the front pages of newspapers and magazines all over the world. The pictures were cuddly but the headlines were mainly full of horror. Journalists imagined that if sheep could be cloned, then so could the worst kind of humans. 'Dreaded possibilities' were raised, according to the *New York Times*. Imagining an army of Hitler clones, *Der Spiegel* declared: 'cloned sheep in Nazi storm'. The *Observer* had broken the story with the warning that: 'it is the prospect of cloning evil dictators that will attract the most attention', and so it proved to be.

'That's never made sense to me,' Ian said. 'I mean, to

be useful, a cloned person would have to be eighteen or twenty years [old].' Should an evil dictator choose to clone himself or herself, they would have to wait a long time before they would have any useful recruits to their 'cause'. An army of cloned toddlers wouldn't be much use. 'If you made a lot of effort to make a cloned army, by the time they'd grown up enough to be useful, the whole political regime would probably have changed.'

Bizarrely (but not unusually, given the double-edged nature of scientific progress), we seemed to be both horrified and delighted by Dolly. She fuelled fears of a *Boys from Brazil*-style future and, at the same time, she was invited to open shopping malls. Anti-monarchy campaigners in Scotland elected her to be their preferred queen. 'She caught the attention of politicians all the way round the world,' Ian said. Dolly the sheep became world-renowned and Ian enjoyed some celebrity status too. He gave countless interviews explaining what he and the team had done and was proud of what he had achieved, scientifically. It was disappointing, therefore, to discover that the science behind Dolly barely got a mention. All the hard work and clever tricks that had gone into creating the world's first clone of an adult sheep were apparently of little interest. Ian himself, however, was considered interesting. *Time* magazine was struck by his appearance: 'One doesn't expect Dr Frankenstein to show up in a wool sweater, baggy parka and with a soft British accent.'

'Given your success as a scientist, I think people will be surprised to learn that you didn't do very well at school . . .' Jim said.

'This is true. In fact, it's a euphemism. I did really badly,' Ian said with disarming honesty.

'How badly is badly?' Jim asked.

'I took three A levels – botany, zoology and chemistry ... but I failed botany, so I only did two out of three A levels.'

He studied agriculture at Nottingham University with a view to doing international agricultural advisory work. 'I don't actually remember it happening,' Ian said, 'but sometime fairly early on in the coursework, I recognised that I'm really not practical and I'm certainly not a businessman. I became increasingly interested in the science, science that I was hearing for the first time.'

'I recognised that I'm really not practical'

To work out if scientific research 'really was a way of life that [he] was interested in' and keen to be near his girlfriend Vivienne (whom he had met at school and who later became his wife), he wrote to laboratories at progressively greater distances from Cheltenham, where she was based, 'asking if they had a gap'. A two-month summer job spent inseminating pigs at the Animal Research Laboratory in Cambridge convinced Ian that a career in scientific research was what he wanted. 'I obviously had to assist with a lot of the mundane tasks,' he said, such as washing up the glassware and surgical instruments, checking to see if the pigs were on heat or not, and keeping records. He also had the opportunity to go to all of the staff meetings and chat to people who were doing scientific research. It was, he decided, 'a really exciting way of life'.

Despite not having performed very well in his A levels, he graduated with a good second-class honours degree and wrote to the Animal Research Laboratory, asking for a reference to help him with his job search. Most unexpectedly, the pioneering scientist Chris Polge wrote back offering him the chance to do a PhD funded by the Pig Industry Development Authority at his laboratory in Cambridge. The person who had been lined up to do the research had committed suicide and there was a vacancy that needed to be filled.

Chris Polge had worked out in the 1950s how to freeze bulls' sperm so that it could be used at a later date to artificially inseminate cows. This innovation allowed cattle breeders to make use of every sperm ejaculated by a prize bull. Previously, many had been wasted. But annoyingly for the pig industry, the freezing techniques that had worked for cattle didn't work for pigs. 'The [defrosted pig] semen were motile, but they didn't achieve fertilisation,' Ian said. His job was to find a way of freezing pig semen so that they remained potent when defrosted. No one understood why the methods used for bulls didn't work for pigs, and finding out why was mainly a question of enlightened trial and error. Ian experimented with changing all sorts of different variables. The rate at which the semen was cooled, for example. It was important to stop ice crystals from forming, because they damaged the cellular structure of the sperm cells. Was this best achieved by flash freezing (dropping the temperature by 100 degrees Celsius in less than a minute), or by a much more gradual cooling process (lowering the temperature by just a fraction of a degree every minute)? Using glycerol as an antifreeze worked well for cattle but not so well in pigs. And there were other factors to consider. Through a series of carefully planned experiments using a piece of equipment that Ian referred to as his 'four-seater loo' (a set of four flasks chilled with liquid nitrogen), he carefully recorded the outcome for every one of his many possibilities, hoping that one set of circumstances might result in frozen pig sperm that remained potent when defrosted.

He finished his PhD, then joined a team that was looking at freezing cattle embryos. 'At that time there was no method for freezing mammalian embryos at all,' Ian said. He joined the group at an exciting time. A few months later they created Frostie and made headlines in the *Daily Mail* in 1974: 'ICE AGE CALF WEIGHS IN'. Frostie was the first calf to be born from a frozen embryo. The technique transformed cattle-breeding programmes. Two years later, using a slightly refined version of the same procedure, the first test-tube baby was created: Louise Brown. The ability to freeze embryos makes IVF much less intrusive. When more eggs are fertilised in the test tube than can sensibly be transferred at one time, the surplus embryos can be frozen and thawed for use at a later date. This reduces the number of times eggs need to be extracted from the mother-to-be. It has also encouraged some women to freeze their eggs in an attempt to beat the biological clock.

Frostie and Louise Brown were the proof of principle. Mammalian embryos could be frozen and could, in some cases, give birth to healthy offspring. But the success rate was very low and the technique still needed to be perfected. For eight years after Frostie, Ian continued to investigate why some frozen embryos survived (when defrosted), when so many others died or grew into animals with deformities, moving to the lowlands of Scotland to work at the Roslin Institute soon after Frostie was born.

The Roslin Institute was part of the Animal Breeding

Research Organisation (a government-funded research organisation that combined practical farming with research to improve livestock production) that specialised in poultry and farm mammals. Animal breeding experiments had been conducted at ABRO since the 1940s. In the 1980s 'the rest of the world was moving on'. A golden age of molecular biology was in full swing and ABRO was criticised for failing to keep up with the times. With the possibility of genetically modified animals on the horizon, the breeding programmes that had created Frostie suddenly seemed old-fashioned and unenlightened. And much of the funding for ABRO was cut.

A new director, Roger Land, appointed in 1982, 'was very keen to introduce molecular genetics into animal breeding'. The plan was 'to look at genes – which genes were associated with rapid growth, or good health, or whatever it was', and then modify an animal's genetic make-up to create farm animals with the most desirable characteristics, and so make livestock farming more efficient. (Such thoughts turned out to be wildly ambitious.) There was also a great deal of interest in another pioneering enterprise: pharming. The idea here was to use animals to manufacture pharmaceuticals in their milk. Tracy, the first transgenic sheep, produced milk that was full of the human protein AAT, which is used in the treatment of cystic fibrosis. Hopes were raised that therapeutic clonings could be manufactured in this way, but Tracy was an exception. The techniques that were used to create her were desperately inefficient and it soon became clear that pharming was not going to be a profitable business.

To make way for this brave new world, Ian was told his research on embryos was 'going to have to stop'. It was receiving 'some scientific acclaim', so 'you can understand that I was exceedingly angry,' Ian said, still sounding bitter. 'Just to give a little bit of background,' he continued, 'normally even moderately senior research workers expect to be able to decide their own priorities, and to select the area of research they want to be involved in.' Unwillingly, Ian redirected his research efforts in accordance with the new director's wishes and started the research work that ultimately led to the creation of Dolly. He had been reluctant to embrace the new science of genetics but, he admitted to Jim, actually 'things worked out really well.'

It all began one night in Dublin. As a gesture of goodwill and perhaps hoping to get Ian enthused, the new director had taken Ian with him to the International Embryo Transfer Society annual conference. One evening in the bar, Ian met an old acquaintance from Cambridge, who was now working for a company and supervising the work done by a very distinguished Danish scientist, Steen Willadsen. Ian's old colleague told him 'that Steen had successfully cloned sheep, and then cattle, from embryos which had twenty or thirty cells.'

'Was Steen Willadsen's sheep the first mammal to be cloned?' Jim asked, surprised to learn of its existence.

'Yes. Yes.'

'So why isn't this sheep better known?'

'I think that's a very good question, not only from the point of view of a general audience, but also from the point of view of the professional groups. He probably doesn't receive the recognition that he deserves.'

'Maybe because he didn't give his sheep a name.'

'Maybe.'

Also, perhaps, because companies don't tend to publish

findings that could lead to others having a commercial advantage.

}

Excited by the news of Steen Willadsen's triumph (and getting in line with the new director's desire to investigate the possibility of genetically modifying animals), Ian decided to see if he could replicate Steen's results, based on what he had learnt in Dublin. He decided to try and clone a sheep using embryonic cells that were just a little bit older than the ones Steen Willadsen had used (nine days, not six).

'What was different about Dolly [compared to the sheep that Steen had created]?' Jim asked.

'Dolly was the first clone from an adult animal.' This is useful for agriculture. If you want to clone your prize bull, you need to be able to first spot the adults with the most potential. Not all their qualities are visible straight away. The behaviour of adolescents is not always a reliable guide to the adults they might become.

In the wake of John Gurdon's success in cloning a frog in 1962,[1] a Princeton professor, Lee Silver, who had spent a lot of time trying to clone a mouse using a similar technique, declared that 'cloning of mammals by simple nuclear transfer is biologically impossible.' If he couldn't make it work with mice, then surely nobody else could, since mice were widely assumed (wrongly, as it turned out) to be the most straightforward mammals to clone. The first mouse was cloned in 2007.

Mice were a model animal for mammalian genetics research, but ABRO scientists were more focused on the end goal. They worked on the animals they intended to breed.

---

1 See the chapter on John Gurdon (p. 95).

The real prize was cattle but sheep were a good place to start. They were much cheaper and more biddable than cows. Nonetheless, cloning an adult sheep was a big step up from adolescent tadpoles. Sheep eggs are thousands of times smaller than the frogspawn used in John Gurdon's pioneering experiments. And they are opaque, which adds another level of complexity, not least that it's very difficult to see where the nucleus is.

'This is probably covered in GCSE biology, but let me check I've got this right?' Jim said, keen to clarify that he understood the principles of nuclear transfer.

'There are three sheep involved,' replied Ian. Three female sheep. One mother provides the genetic material so that it can be cloned. Another mother donates embryonic cells to provide a home for these nuclei. And a third mother acts as a more conventional surrogate mother, providing a womb in which these genetically modified cells can grow. Cloned sheep do not have fathers. (Rams are introduced to detect whether the ewes are on heat, but that's about it.) Males are 'replaced by an electric shock!' Ian joked. 'That's what we do to the egg instead of providing sperm . . . That's what tricks the egg into thinking, "OK, I should begin to develop now."' Then, with any luck, the cells start to divide and multiply, just as they would have done had they been fertilised by sperm.

So much for the theory. Researchers at ABRO were used to handling eggs, sperm and embryos. They had invented all kinds of techniques for assisted reproduction. They knew how to transplant embryos from A to B and create suitable conditions for them to grow in test tubes and in wombs. But nuclei are significantly smaller than embryos and ABRO scientists had no experience of moving nuclei from one cell to another. Nuclear transfer required complex surgery to be performed on an egg that was barely visible to the naked eye. How do you extract a dense tangle of DNA (no more than 5 billionths of a millimetre across) and introduce it into a de-nucleated cell?

'I think it's important to emphasise, if I may,' Ian interjected, 'that a project like ours is very much a team project with twelve, fifteen people, altogether involved, all the way through, from looking after the animals to growing the cells, micromanipulation and so on. I'm happy to be regarded as the leader of the team.'

'It had been known for a number of years that if you transferred nuclei from one cell to another, it was important to coordinate the cell cycle,' he went on to explain. A lot of changes occur in the run-up to a cell dividing to create two daughter cells. Ian's colleague, Keith Campbell, had studied the cell cycle when he was working on a PhD on cancer cells. He designed a set of experiments to try and work out when it might be a good time to remove and insert nuclei, spurred on by the knowledge that Steen Willadsen had got it to work, so there must be a way. 'Keith . . . certainly anticipated that it would be possible to clone adults, sometime,' Ian said, 'but I don't think that he really believed that we were going to do it in a short series of experiments.'

What Keith discovered was this: it was all about coordinating the timing. Nuclear transfer could be got to work in sheep if the cell cycles of both the donor and the host cells were in sync. The nuclei from the cells of the genetic mother needed to be extracted from cells that were at the same stage in the cell cycle as the recently de-nucleated cells into which they were about to be placed. Furthermore, Keith worked out that, 'There were two stages [in the cell cycle] that appeared to be capable of producing an embryo which would develop normally.'

Fresh, ripe eggs were put in shirt pockets and tucked into bras to keep them warm while they were being transported to the lab from the barn. In the lab, nuclei would be removed using a micromanipulator, and the nuclei from the eggs that had been stored in the freezer were inserted.

Most attempts to clone sheep failed. In the experiment that led to their first success, 244 embryos were created using nuclei that had been extracted from embryonic cells. Only thirty-four were viable. These thirty-four genetically modified embryos were then implanted into surrogate mothers. Many had miscarriages. As one sheep approached full term, Keith's hopes were raised. Desperately concerned for her welfare and keen to do everything he could to ensure the survival of her offspring, he camped out in his office near the barn. Five lambs were born. Only two survived. When these two lambs grew into healthy sheep, the team rejoiced. SLL2 and SLL5 were renamed Megan and Morag, following a suggestion from Keith's wife. And the results were published in *Nature*, with Keith as the first author on the paper. Steen Willadsen's unpublished result was confirmed: sheep could be cloned using embryonic cells. The next step was to see if the nuclear transfer techniques could be made to work with nuclei taken from adult sheep, not nine-day-old embryos. '[Megan and Morag] were the first clones from differentiated cells, and it was that, I think, that led us to be confident that one day, somebody would clone an adult [sheep],' Ian said.

Thrilled by their success with Megan and Morag, Ian and Keith decided there and then to try the same procedure using nuclei that were taken from adult sheep cells. If this was possible, then animal breeders would want to know. Fortunately, the freezer at the Dairy Research Institute was full of breast cells that had been collected from sheep that were six years old and stored, ready to be used in a study on lactation that never happened. One of these long-dead sheep was to become Dolly's genetic mother. Thanks to another research project that never came to fruition, there were also plenty of frozen sheep embryos. The experiments that led to Dolly (and there were hundreds of them) involved removing the nuclei from these embryos and replacing them with nuclei taken from the breast cells of a six-year-old sheep. Then the trick was to kick-start an unfertilised egg into dividing and multiplying, as if it had been fertilised by sperm. The embryos that thrived were then placed in the womb of a third sheep, the surrogate mother.

'Dolly was one out of two hundred and seventy-seven embryos,' Ian reminded Jim. 'Now, roughly speaking, Bill – who did all the micromanipulation – was only successful about half the time. So, if he produced two hundred and seventy-seven eggs that were suitable for transfer in total, there were probably something between five hundred and fifty and six hundred attempts to produce that.'

'And of those hundreds of embryos, only twenty-nine were then implanted into surrogate mothers?' said Jim.

'Yes. That's correct.'

Their gestation was an anxious time for all the scientists involved. They were waiting to see if any of these embryos would grow into healthy lambs.

'Given that these experiments inevitably have a high

failure rate,' Jim said, 'how concerned were you that Dolly might die soon after she was born, or that there might be something wrong with her?'

'Very, very,' Ian replied, definitively.

'I imagine most of the embryos failed in the early stages of pregnancy, but were there many miscarriages in the later stages?' Jim asked.

'There were, and even after birth the postnatal loss rate was increased in ways which we don't think would occur normally. We had one ram lamb who was big, well-built, very energetic, running around very quickly, but it panted all the time. Even if it was just resting, it panted. You cannot detect [abnormalities like this] whilst the lamb is still in the mother's uterus, it's such a small thing. So, you just couldn't be sure.'

'It must have been scientifically disheartening, of course,' Jim said. 'But did you find abnormalities like this personally distressing?'

'I think . . .' Ian started to reply, then paused, for the first time in the interview. 'I think the word I'd use would probably be "disappointing". Perhaps not "distressing", because we, sadly, are used to seeing them in difficult circumstances. But [it was] certainly very disappointing that we weren't yet being more successful.' It was hard to adjust to the reality of another failure after their hopes had been raised by the birth of an apparently healthy lamb.

'I know one study conducted at Roslin concluded that four in ten of cloned lambs died before they were a month old,' Jim reminded Ian. 'That's quite a big number.'

'I think it's important to make the point that embryos die in ordinary reproduction,' Ian replied. 'In fact, our own species is the most vulnerable to losses of this kind.

Embryos are lost even in very healthy people and animals. So, not all of what we're seeing would be due to cloning.'

}

After hundreds of failed attempts, a healthy lamb was born in July 1996. Immediately after her birth, they were concerned that she might not survive. So many lambs had not made it in past experiments, after all. But with every week that passed their excitement grew.

'Why was she called Dolly?' Jim asked.

'The cell that was used to provide the genetic information came from mammary tissue, and she was named Dolly [in reference to] the spectacular mammary tissue that Dolly Parton had. When Dolly Parton's manager learnt that his protégé was now closely associated with a sheep, apparently he said: "There's no such thing as baa-aad publicity."'

'She was named Dolly in reference to the spectacular mammary tissue that Dolly Parton had'

The team were sworn to secrecy about Dolly until their results had been written up, for fear of being trumped by rivals. And, unfortunately, many people at the Roslin Institute first learnt about what their colleagues had achieved in the Sunday papers, when the *Observer* broke the story (ahead of the publication embargo imposed by the institute).

Dolly survived into adulthood but died when she was six years old, having given birth to six children of her own. Sheep usually live to about twelve years old. Dolly's genetic mother (whose breast cells had been stored in the freezer) was six when the cells were taken from her and there was widespread speculation that Dolly's premature death was somehow predetermined by the age of the sheep from which she had been cloned. Many felt that the fact that Dolly had arthritis (which was unusual for a sheep of her age) strengthened their case: the belief that cloned animals would never be as healthy as those that were created in a more conventional way.

But Ian never believed Dolly's unusual start in life was responsible for her premature death. 'There's no reason to think that this was anything to do with cloning,' he said. 'A fully detailed histopathology was done by one of the head veterinarians at the vet school here in Edinburgh, and she found nothing that persuaded her that this was the case, and thought it was much more likely that the effect had been of a cancer, which was nothing whatsoever to do with cloning ... There is a virally induced cancer in sheep which is relatively prevalent in this part of Scotland. It's not very common, but it is a known issue, and there is no cure for it. My veterinary colleagues were shocked at the size of the tumour that they found. So we discussed this and decided in the end that it was kinder, again, to end her life rather than to let her recover, only to ultimately succumb to this terrible disease.'

Keith Campbell moved to the University of Nottingham and set up his own lab. Using breast cells taken from the same genetic mother as Dolly (there were more in the freezer), he set about trying to produce more clones, hoping to test the hypothesis that Dolly's premature death was because she was cloned from a six-year-old sheep. Daisy, Debbie, Denise and Diane were born in 2007 and they all survived beyond the age of six, 'proving quite categorically that that isn't the case, that the clock is reset,' said Ian. The longevity of Dolly's sisters proved that clones are not predetermined to die prematurely. 'They are full sisters of Dolly, and they haven't exhibited any of [her] abnormalities.'

}

In the wake of Dolly, cattle, horses, mules, oxen and rabbits have all been cloned. A decade after Dolly, scientists also cloned a rhesus monkey. Aware of the trend towards cloning ever more sophisticated animals and failed attempts by South Korean scientists to clone *Homo sapiens*, Jim felt compelled to ask: 'Do you think that one day someone might be able to clone humans?'

'I think we probably could clone humans now,' Ian said. 'I think it could be done.'

'You think it could be technically possible in your lifetime?' Jim asked.

'Yes,' Ian replied. 'During the last year or two, two groups – one in the West Coast of the United States, one in New York – have worked together and looked at the process of cloning in humans and other primates. What they found, first of all, was that there is something very different about this group of animals. These groups investigated the reason why, and it does seem as if it could be done now . . . What I would question is the reason why people want to do it, because I don't think there is a reason which is appropriate, either because of the effect on the cloned child or because there is another way of achieving the same objective.'

The reproductive cloning of humans is banned in many countries, but not all. It is legal in South Korea, for example. 'I don't think you could ever stop some maverick scientist, in some country with lax regulations, to go ahead and do this,' Jim said.

'It could become a crime against humanity,' Ian suggested. '[Then] it wouldn't matter where the person did the work.' Offenders could be tried in the court in the Hague.

'I have to say I agree with you there,' said Jim.

Dreaded possibilities are indeed raised by cloning mammals, as *Der Spiegel* predicted when journalists first learnt about Dolly. But, as is so often the case when science advances, the new knowledge generated can be used in many different ways: as a force for bad or good, or a messy mixture of the two.

The European Parliament banned the cloning of farm animals in 2015. It remains legal in the USA and China, where there are plans to build factories capable of producing thousands of clones of highly prized meat-producing individuals and breeds, thereby providing top-quality beef more cheaply. (The Food and Drug Administration in the USA declared beef from cloned animals to be as safe as meat from more conventionally farmed animals in 2008, and some food critics suggested that it was equally delicious.)

Meantime the global market for cloned pets is thriving. Companies based in South Korea and in Texas now offer to prolong the memory of much-loved cats and dogs. Copycat, the first cloned pet, was created in 2001. Unfortunately for her owner, (who had paid a small fortune, hoping to replicate her beloved pet) Copycat was not an exact copy of her mother. In fact, she looked nothing like her and behaved quite differently. Clones share exactly the same genetic material but their appearances may differ, due to epigenetic effects. Not everything about us, or our pets, is determined by our genes.

Cloning pet animals is part of Dolly's legacy, as is the industrial manufacture of cloned beef. Therapeutic cloning could help to save lives. Reproductive cloning is banned in Britain but therapeutic cloning is actively supported. The nuclear transfer technique that Ian and Keith and the team at the Roslin Institute perfected in order to create Dolly is now widely used to clone stem cells for medical research. And in 2003, Ian joined the Medical Department at Edinburgh University to work on therapeutic cloning. 'Stem cells are being used to study inherited disease,' he said. 'If we have somebody who we know has inherited motor neurone disease, we take some of their skin and make stem cells, using therapeutic cloning techniques. These pluripotent stem cells can then be grown up into whatever cells you want to study: the nerves which are damaged in motor neurone disease, for example.

'Scientists are now able to study [cells] that have been obtained from patients and cloned.' These cloned cells, taken from someone who is unwell, can then be compared with cells that have been obtained from a healthy counterpart. 'I have colleagues in the centre here in Edinburgh who have done that and found differences,' Ian said. 'We're beginning now to understand, for the first time, what is different about the cells of patients who have motor neurone disease.' A detailed understanding of how a condition like motor neurone disease progresses at a cellular level could transform our ability to treat this disabling condition. 'Dozens of diseases could be studied in this way,' Ian said.

The ability to create cells that are genetically identical to the patient's own is transforming medical research. 'Dozens of diseases could be studied in this way.' And this, Ian believes, 'is perhaps the greatest inheritance of the experiments that created Dolly the sheep.'

ROBERT MAIR

'The ground is full of surprises'

Grew up in: Cambridge
Home life: married to Margaret with two children, Julia and Patrick
Occupation: civil engineer
Job title: Professor and Head of Civil Engineering at Cambridge University
Inspiration: Sir John Baker and his collapsing steel table
Passion: constructing tunnels under busy cities
Favourite invention: compensation grouting to prevent buildings from being destabilised by tunnelling
Mission: to build tunnels through soft ground
Advice to young engineers: 'Never stop challenging the status quo'
Date of broadcast: 15 January 2013

Professor Lord Robert Mair designs and builds tunnels. As a young man he became interested in soil mechanics and, after working for a big engineering consultancy company for a number of years, he set up the Geotechnical Consulting Group, offering expert advice on all aspects of geotechnical engineering, and specialising in tunnelling through soft ground. He has worked on metro systems in Hong Kong, Baghdad and Singapore (where the ground had the consistency of toothpaste), the Channel Tunnel, Crossrail and the Jubilee Line Extension for London Underground. He invented compensation grouting to minimise ground movements resulting from large-scale underground excavations and prevented Big Ben from leaning like the Tower of Pisa. The system he created for testing the feasibility of tunnels is used all over the world. After twenty-seven years in industry, he is now a professor of civil engineering and (at Cambridge University) in charge of several multimillion-pound initiatives to improve the links between industry and academia and to encourage innovation.

Jim Al-Khalili interviewed Robert in January 2013, when Crossrail engineers were tunnelling under Central London.

'I was in a tunnel a couple of days ago in London as part of the Crossrail project, and I have to say I still find it utterly fascinating and all-absorbing,' Robert told Jim.

Mindful of all the tunnels that now exist underneath London, Jim remarked: 'I do have this vision of London now sitting on a honeycomb of mostly empty space.'

'There isn't much soil left under Waterloo station,' Robert confirmed. 'It's all tunnels. But it is, I have to say, safely constructed.'

'Because these tunnels are all lined with this strong concrete?'

'Yes, that's right.'

> 'There isn't much soil left under Waterloo station. It's all tunnels. But it is, I have to say, safely constructed'

'[Tunnelling] throws up all sorts of interesting challenges,' Robert said. 'In many other forms of engineering, one's dealing with steel or concrete, or silicon chips, or whatever it might be, but in the case of tunnelling, one is dealing with whatever nature has provided.'

Robert specialises in building tunnels through the soft ground that is found under most cities. 'And the soil is never uniform. Over the previous hundreds of thousands of years, there have been tricks played, changes going on, streams or old rivers that one didn't know about.' Ancient rivers have changed the nature of the surrounding rock. The geology of the area may be quite well understood and generally quite solid. But often there are unexpected seams of weaker layers (soft clays or loose sands, for example), on the sites of ancient river beds, which can cause major problems.

'No matter how many boreholes one might do before the investigation, it's quite common to miss something. You're always having to be on your guard as the tunnel is being constructed. The ground will not be one hundred per cent as you expected, and the art of doing a really good design is to try to anticipate all that.'

Immersed in science from a very young age, Robert always enjoyed thinking about how things worked. It was the inventor and distinguished engineer Sir John Baker who piqued his interest in the structure and properties of materials. Aged sixteen, Robert invited him to give a talk to his school science society.

'I think one has these moments in life when certain people have a very strong influence,' Robert said. John Baker, who was quite a showman, arrived with a model of the indoor shelter he had invented to protect families during bombing raids in World War II. The full-scale version was a steel structure that could double up as a kitchen table by day and was big enough for a family of four to sleep under when the air-raid warnings sounded at night. 'With a great flourish, [John] got out his gold watch from his waistcoat and he put his gold watch under his model.' All eyes were fixed on the very large weight that loomed over this relatively flimsy-looking table with wire-mesh sides. 'Then he dropped this huge weight on to this structure, with a great crash.' The structure deformed. It crumpled and buckled but – miraculously, it seemed – it was not completely crushed. And the gold watch remained intact.

'That was a formative moment,' Robert said. When John Baker explained how it worked, he listened, as if to a magician explaining how he had done a trick. Keen to be similarly empowered, he decided to study engineering and went to Cambridge University. His father, a professor of aeronautical engineering, was 'neutral but encouraging' about his choice of degree subject, but John Baker's performance had convinced Robert.

A lecture by Peter Wroth (author of a landmark paper, 'On the Yielding of Soils', written in 1958) led him to specialise in soil mechanics. Peter described how different soil structures reacted to external forces, and in the wake of the Aberfan coal-tip disaster in 1966, this felt important.

This disaster had happened when Robert was sixteen, and the public inquiry was still going on by the time he was a student. 'Do you remember your reaction to that disaster when you heard about it in the news?' Jim asked.

'I can remember that very vividly. I can remember the awful news of one hundred and sixteen schoolchildren being killed, and a large number of adults too. And I can remember that there was much discussion in the press about why and how it had happened.'

Tip No. 7, a vast heap of waste and tailings that had been dumped at the top of the valley just above the village of Aberfan, blocking the mouth of a mountain spring, slipped down the mountain. The spring water was unable to drain away and a pile of solid waste 34 metres high turned into an unstable heap of thick, black sand. Expanding as it absorbed more and more spring water, it spilled over itself and rose up. On the morning of 21

October 1966, an avalanche of black waste started to descend into the valley, falling over itself in waves. It hit the mains water supply that connected Cardiff to the Brecon Beacons and the pipe burst. Mains water was injected with some force into this monstrous mass of liquefied coal waste, which grew bigger and bigger and flowed downhill ever faster, sometimes reaching speeds of up to 40 miles per hour. Tragically, 116 children and 28 adults were killed while preparing for morning assembly at the local school in Pantglas.

As the full details of what had happened emerged during the public inquiry, there was 'a lot of talk about liquefaction', Robert said. People were trying to understand 'how it could happen ... how particulate material can change character so quickly and unexpectedly: from being an absolutely stable pile of waste material from a coal mine to suddenly flowing at a speed of thirty or forty miles an hour, straight down the mountainside into the village, engulfing the school.' It was a dramatic and terrible reminder of how the properties of materials can change. And to get to the bottom of what happened in Aberfan, an understanding of soil mechanics was required. '[The Aberfan disaster] became a seminal moment for the science of soil mechanics,' Robert said.

'The Aberfan disaster became a seminal moment for the science of soil mechanics'

He joined the civil engineering consulting company Scott Wilson Kirkpatrick (as a graduate trainee) because they were doing interesting geotechnical work, and he was sent to Hong Kong, aged twenty-three, to help design and manage construction of a huge new container terminal. There was a construction boom and Hong Kong was an exciting place to be. As soon as a project got the go-ahead, things happened very quickly.

'[In the early 1970s] it was becoming very fashionable to think about a lot of tunnelling schemes,' Robert said. 'The big challenge was: nobody really knew what happened if you created the tunnel in soft ground, in and amongst and beneath buildings, particularly in a place like Hong Kong, which already had lots of high-rise buildings on foundations.'

The Hong Kong government was keen to build a metro system and Robert was involved in preliminary ground investigations. Meanwhile in the UK, ground engineers at the Transport and Road Research Laboratory were wondering if it would be feasible to build a tunnel between England and France.

'And so Scott Wilson in Hong Kong, combined with the UK government, sent me back to Cambridge to do a PhD on that subject: is it safe to make a hole in the ground when the ground is soft? What movements may take place? Would it be unstable? Might the tunnel collapse? And all those things.'

'Is it safe to make a hole in the ground when the ground is soft?'

To try to answer these questions, he decided to create a scaled-down model of a tunnel in soft ground and use it to predict how tunnels in different soils might behave under different conditions. If aeronautical engineers could test the performance of their planes by building scaled-down models of their designs and subjecting them to different wind conditions in manufactured wind tunnels, maybe he could do something similar to find out how the surrounding ground would respond to the new forces that would be acting on it, when a tunnel was introduced? At what point would the soil structure collapse? Typically, soils (which are made from a delicate matrix of earth, water and air in varying proportions) have a non-linear response to stress. The ground will distort and compress until it collapses. (The Aberfan landslide is a good example of this.)

Making scaled models was not a new idea but modelling ground conditions introduced a particular problem: how to accurately represent gravity. Ground strength was known to be highly sensitive to confining pressure (typically caused by the weight of the soil), so it was an important variable. It was difficult to model, however, because gravitational forces are dependent on the mass of an object. The gravitational force is the mass of an object multiplied by the acceleration due to gravity, which on Earth is a constant (9.8 metres per second squared). So, if for example, Robert built a model that was one hundredth of the size of the real thing, then the acceleration due to gravity would need to be one hundred times what it is on Earth, to achieve a realistic representation of the gravitational forces that would be at work in the real world. The solution he came up with was ingenious. He put his model of the tunnel and ground conditions into the newly constructed 10-metre-diameter centrifuge at Cambridge and spun it around at high speeds, and so created an enhanced gravitational field of 100 G.

Scaled-down models of different tunnel designs could be put in this spinning, gravity-enhancing machine and used to predict how precisely the earth might behave. It opened up the possibility of testing a range of tunnel designs in varied ground conditions and enabled engineers to calculate likely ground movements.

The centrifuge modelling system that Robert used for his PhD was the first of its kind and is now used by industries all over the world. He finished his PhD in 1979 and, after working for Scott Wilson Kirkpatrick for another few years (designing a metro for Baghdad, among other things), he decided to set up his own company, mainly to avoid being 'sucked into senior management roles'. He wanted to stay close to the technical side. Robert and David Hight (an old friend from SWK who had since become a professor at Imperial College) co-founded the Geotechnical Consulting Group in 1983, offering specialist advice on soil mechanics to construction companies on many different projects, including those who were building tunnels in urban areas, using Robert's modelling system to guide their decisions and making the most of the latest academic research on soil mechanics.

Sitting in a small office above a car showroom in a mews house in South Kensington (conveniently close to Imperial College), Robert wrote reports for construction companies and offered expert advice and, all the while, continued checking the accuracy of his modelling system whenever he could. 'Robert kept a very tidy desk,' David said. 'But always on his desk was a piece of graph paper.' He would collect data avidly on ground measurements that had been taken when tunnels were being constructed, reading 'anything he could lay his hands on but particularly anything about tunnels under London'. And then he would plot his findings on a graph in order to be able to see at a glance how what was happening in the real world compared with what his spinning model would have predicted. 'Sometimes the excitement was such that it meant bringing me up from the basement to look at his newly plotted point,' David said. 'So, I would go up and admire the point. And [Robert] was particularly delighted if they fitted within his predictions, which they often did.'

There was no shortage of problems for GCG to solve. Rapidly growing populations in built-up and congested cities were forcing transport planners to dig deep in ground conditions that were often far from ideal. 'Most cities in the world are built either on rivers or on the coast,' Robert explained. 'There will eventually be rock if you go deep enough, but the geology beneath most cities is usually pretty soft soil.'

London is built on soil. The chalk it rests on lies much deeper. Singapore is built on about 30 or 40 metres of very soft clay, 'which is about as strong as toothpaste'. 'It's probably one of the most challenging types of soils to create tunnels in, because if you try to make a hole in toothpaste . . .'

'Is it possible?' Jim asked, wide-eyed.

'We now know how to engineer that,' Robert replied, 'and we know how to do the necessary calculations to make sure that it's done safely. But the subject of my PhD back at Cambridge was exactly to evaluate that. For a ground of given strength, how large can the tunnel be? How deep can it be? And what are the consequences?'

'For a ground of given strength, how large can the tunnel be?'

When the minister for transport in Singapore wanted to push ahead with 'the most expensive single project undertaken in Singapore', Robert did the calculations using the centrifuge modelling system he had developed during his PhD. Ground conditions were about as bad as they could be, and to avoid all the pre-existing underground tunnels (pedestrian walkways, service tunnels for cable and roads), they needed to be more than 30 metres deep, which made matters worse. Nonetheless, plans for a mass transit system went ahead and Robert's expertise was very much in demand. He was later awarded a Singapore Public Service Award for all his invaluable advice on soil mechanics and tunnel construction.

'The work done by GCG is known throughout the world,' Jim said. 'You've been involved with so many different projects. The Channel Tunnel rail link, the extended Jubilee Line for London Underground, building the longest escalator in London at Angel station, advising on collapsed tunnels in Turkey and France . . . Do you have a favourite project?'

'Ask me in a few years and I might say Crossrail, but for now I think it's probably the Jubilee Line Extension,' Robert said. 'It was a superbly challenging and interesting project.' For five years, he worked on very little else, figuring out how to connect Westminster, South London and Canary Wharf.[1] Innovative engineering solutions were required to tunnel at depths of 30 metres, through difficult water-logged sands and gravel in South-east London. The new line crosses the Thames four times. Five stations

---

1 Canary Wharf was the fastest-growing central business district in Europe at the time. The Docklands Light Railway couldn't cope with ever-increasing numbers of commuters travelling to the Isle of Dogs. And everything needed to be ready in time for the planned celebrations at the Millennium Dome in Greenwich on New Year's Eve 1999.

were substantially rebuilt and six brand-new ones were created, including a state-of-the-art very deep station at Westminster.

Geotechnically (as well as politically) it was an exceptionally sensitive location. The home of British democracy was built on soggy London clay by Victorian engineers who were rather more gung-ho than their modern counterparts about establishing firm foundations. 'The Victorians simply dug a hole in the ground and carried on digging down until the water was coming in and it was time to stop. And they then filled it up with what would have been, in those days, quite primitive mass concrete. And that was the foundation of Big Ben ... And sometimes the pressure, the stress, acting on the foundation is really rather high. Higher than most modern engineers would specify.'

Even before they started drilling holes and hollowing out great long tubes, Big Ben was at risk of going the same way as the Leaning Tower of Pisa. The plans for the new Jubilee Line station at Westminster would make it the deepest station in Europe, and it would require a lot of very deep tunnels to be built not far away. Deep excavations in soft ground inevitably cause the surrounding ground to move. The magnitude of ground movements is directly proportional to the depth of the excavation. Typically, these movements extend outwards from the excavation site to a distance that's about twice the depth of the hole that has been created. Building tunnels that were 30 metres deep, 28 metres from Big Ben was, therefore, a definite cause for concern, and monitoring the tilt of Big Ben was one of the most important parameters when the new Westminster station was being built.

When a new five-storey underground car park was

built in the nearby New Palace Yard to provide parking spaces for MPs, the surrounding ground had shifted and Big Ben had tilted, leaning a little to the left when viewed from Parliament Square.[2] The excavations that would be needed to create the new Jubilee Line station were on a much larger scale. (The tunnels would be the equivalent of fourteen storeys deep, not five and the main station would be one of the biggest in London.) It was clear that protective measures would need to be put in place to avoid Big Ben tilting significantly more than it had done when the car park was being built.

'What could have been the worst-case scenario?' asked Jim.

'I think [Big Ben] would have leant to such a degree that it would have caused unacceptable cracking.' A split would have opened up between Big Ben and the Houses of Parliament. 'Big Ben is actually linked to the Palace of Westminster, and all of that would have been quite seriously damaged, as Big Ben tried to lean to the north.'

To solve this problem, it was decided to use a new technique that Robert had invented and tested (very cautiously) when the new Eurostar terminal was being built at Waterloo station. By injecting a liquid grout in the ground above the tunnel, he planned that any ground movements that came about as a result of the tunnelling would be intercepted by the grouting and would not reach the buildings on the surface. He had put his compensation grouting to the test (introducing tiny changes and taking lots of measurements at every stage) and had found that it worked well. Emboldened, he then decided

---

2 Thankfully, perhaps, politicians who leaned to the right were unable to detect this shift.

that it was safe to use compensation grouting to protect Big Ben.

'A shaft was dug in the middle of Bridge Street, which runs up to Big Ben. During the daytime the opening to the shaft was decked over with a steel plate, so the traffic could carry on flowing. At night-time it was opened up and steel tubes were drilled deep down in the ground, horizontally, right under the foundations of Big Ben.' An array of pipes lay underground, spreading outwards from Big Ben like a fan, and each with a series of holes at regular intervals along its length which were covered with rubber flaps. Before each drilling operation took place, the precise location and nature of the ground movements that would result were predicted. Armed with this knowledge of how and where the ground was going to move, the nearest hole could be selected, and grout could be delivered to precisely where it was most needed.

'It was precision engineering . . . As the tunnelling was taking place, and the movements of the ground were happening, we would compensate for those movements by injecting very well-defined, small quantities of liquid cement.'

'So the idea was that you could continuously monitor how the ground might be changing and immediately do something about it?' Jim asked.

'Yes. It relies very heavily on having instrumentation on Big Ben constantly feeding information to the engineers. Every few hours a decision was being made. How much tunnelling had taken place? How much tunnelling was likely to be taking place in the next four hours? Which tubes should be injected? And each of those steel tubes, I should say, has holes in it, typically about six hundred millimetres apart. And you can choose one of those holes to inject the cement from – the liquid cement.'

Fifty litres of liquid cement were injected into different specific underground locations on twenty-four separate occasions. 'That was it!' Robert said, clearly still pleased that his simple and innovative solution had worked. The 12,000 litres of compensation grouting were enough to make sure Big Ben 'did not move more than very small specified amounts'.

'It has tilted quite a bit further since construction stopped in the late nineties,' Jim reminded Robert. 'Is that something that's related to the Jubilee Line Extension?'

'We always knew that there would be subsequent movement of Big Ben,' he said. 'We predicted that. We knew that [Big Ben] would go on leaning to the north,' as the London clay beneath it accommodated new stresses and strains, swelling and consolidating as the water pressure within it adjusted to the new regime. 'That was all monitored very carefully and it was all leaning to the north within our predictions. And so we were happy about that.'

Nonetheless, the clockface has now moved a total of 35 millimetres. 'Measurements are still being made, long after completion of the project, just to keep an eye on what Big Ben's doing,' Robert said. 'There is always the possibility, if needs be, even today, to go back and do more grouting. If, for some reason, Big Ben had moved more than what's expected, or had got to the point when it was necessary to do that. But it has never reached that point.'

> 'Measurements are still being made, long after completion of the project, just to keep an eye on what Big Ben's doing'

The demand for tunnels under London remains high. The tunnels needed for Crossrail were considerably bigger than any of the London Underground tunnels that had been built before. Jim's interview with Robert came midway through the construction of the Central London section.

'Probably the most important point about Crossrail is that the tunnels are much larger in diameter than existing Underground tunnels,' Robert said. 'A typical London Underground tunnel between stations is about five metres in diameter. The Crossrail tunnels are about seven metres in diameter. That may not sound very much, but it's fifty per cent bigger in terms of things like settlements.[3] Ground movements are proportional to the square of the diameter. So, if you're going up by fifty per cent, then you're more than doubling the likely settlements.' The risk of existing structures sinking into the ground increases rapidly as the diameter of the tunnel increases.

'As we speak, a tunnel is being excavated – not quite under our feet here at Broadcasting House[4] – but it's very close by. A tunnel-boring machine is heading towards us . . .' said Jim. 'And this particular one, I gather, is called Phyllis! Do they all have names?'

'They all have female names,' said Robert. 'I'm sure there's a historical reason for that.' Phyllis was named after Phyllis Pearsall, who created the *London A–Z*. Another one is called Ada after Ada Lovelace, the computing pioneer. Phyllis and Ada were each boring their way through 6.8 kilometres from Royal Oak to Farringdon,

---

3 A polite technical term for sinking.
4 In London W1A.

creating twin tunnels in their wake. Working twenty-four hours a day, seven days a week, it had taken Phyllis eight months to make it to Oxford Circus – about halfway.

A hefty sausage-shaped machine, 7 metres in diameter (the same width as the hole required) and 150 metres long, she weighs 980 tonnes. Powered by thrust cylinders that push off against concrete lining segments, she slowly moves forward, leaving a smooth-walled tunnel behind her. A rotating cutting wheel on her nose loosens the ground ahead and a conveyor belt at the back removes the excavated earth. Twelve members of the tunnel gang work inside the machine on 12-hour shifts. There's a kitchen and a toilet on board.

'It's a serious bit of machinery and a marvellous example of how tunnelling technology has advanced so much in the last few decades,' Robert said. 'If you make a hole in the ground, there is a potential for instability. There's a potential for that hole to move or, at worst, even collapse. These modern tunnelling machines can now be made to pressurise the face, so that as the big wheel's turning around and cutting the soil, there's a fluid exerting a continuous pressure on the soil ahead. And what that means is that it can reduce the amount of movement to very small values. What is really exciting is the fact that the magnitude of the ground movements that we are now seeing on a project like Crossrail is significantly less than it would have been

'What is really exciting is the fact that the magnitude of the ground movements that we are now seeing on a project like Crossrail is significantly less than it would have been on the same project perhaps ten or twenty years ago'

on the same project perhaps ten or twenty years ago.'

'So this pressure it exerts compensates for any movement that might take place as it's removing the soil?' Jim said.

'Yes. And the holy grail is that if you can have a pressure exerted on the tunnel face that is equivalent to what was there in the ground before the tunnelling, you're not going to move it at all! It may never actually quite look like this, but there is the prospect of being able to do the tunnelling with zero movement.'

After twenty-seven years working in industry, Robert was drawn back into academia in 1998. Alec Broers, the then vice chancellor of Cambridge University (himself an engineer), decided that Robert's industrial expertise and broad experience of civil engineering projects worldwide were just what Cambridge needed, and invited Robert to become professor of geotechnical engineering and head of civil engineering.

'It was a difficult decision for me. Not just for me, of course, but also for Margaret, my wife. We were living in London, our children were at school in London, Margaret had an important job in London . . . and so it was a difficult decision, but I'm glad [Alec] persuaded me.'

'Do you feel that engineering has its rightful place in people's perception of broader science?' asked Jim.

'I think the standard view of engineering in the eyes of much of the general public is one of heavy machinery and dirt and grime, and [the] rather large-scale heavy end of engineering, which of course is only one part of it. And, of course, when people ring up and say their fridge is broken, they're very often told that the person who comes round to fix it is an engineer. So that's the sort of thing we're constantly battling against in this country.'

Alec wanted to forge stronger links between industry and academia. 'Do you think that's what's needed to increase the profile and number of engineers in general?' Jim asked.

'I think that's very important. The need to have very close collaborations between industry and academia is

'I think the standard view of engineering in the eyes of much of the general public is one of heavy machinery and dirt and grime, and [the] rather large-scale heavy end of engineering, which of course is only one part of it'

paramount, and I think this need is increasingly being recognised in many branches of engineering. I sometimes think there's an analogy with medicine. The top professor in medicine at a university will nearly always be involved with patients. And that's absolutely right. And I think the same should be said about most top professors in engineering.' They should be applying their minds to some of the difficulties that inevitably arise when the technologies they have invented are applied in the real world.

'Crossrail is a living laboratory for us at Cambridge,' Robert said. 'We have quite a number of PhD students who are down in the tunnels in Crossrail, making measurements, putting optical fibre into linings, experimenting with wireless sensors.'

In 2011, Robert became head of the new Innovation and Knowledge Centre for the Cambridge Centre for Smart Infrastructure and Construction, funded by the government and industry, including many of the UK's leading construction companies. CSIC is a £17 million project which involves Cambridge academics collaborating with thirty different companies to monitor the behaviour of buildings, bridges, tunnels and other types of infrastructure by installing remote sensors.

'There's been a revolution in sensors,' Robert said. And there's been a revolution in wireless technologies. So the notion behind this very large research project is to use sensors in a completely novel way. During actual construction, sensors can be embedded in piles, in bridges, in tunnels – partly to make more efficient the actual construction process itself. If you measure things, you know exactly what's happening and you 'can make decisions in a much more rational way'. The sensors can also be used once the project is complete. The construction company can say to the owner: 'This is how your bridge, tunnel, flood defence, whatever it may be, is performing.' 'And that,' Robert said, 'can be extremely valuable for the whole way in which its maintenance is covered, throughout its design life. You get a much better picture of how infrastructure is performing.'

'You can put a life span on buildings and bridges . . .?'

'Yes. Often nowadays the question is asked: "How much longer has our piece of infrastructure got?" In autumn 2012, London Underground was asked by financial institutions: "Is the Northern Line good for another twenty years? Or for another fifty years or a hundred years?" And the answer is: "Put sensors into it, and then we can tell you the answer."

'The construction industry in some senses has lagged behind other forms of engineering. If one thinks of the aerospace industry, or the automotive industry, there is a widespread use of very sophisticated sensors. As an example, Rolls-Royce engineers in Derby can monitor exactly how their engines are performing anywhere in the world. They can see how their engines are behaving, and whether one needs to have a new fan blade, for example. We do very little of that in the construction industry.'

In future, perhaps, our infrastructure and buildings will be able to communicate with the men and women who designed and built them, alerting them to new stresses and strains, and warning us all when they need attention or feel unsafe.

ANN DOWLING

'Flight has always seemed quite incredible to me'

Grew up in: born in Somerset, lived in Egypt, Singapore and various places around England
Home life: married to engineer Tom Hynes
Occupation: mechanical engineer
Job title: Professor of Mechanical Engineering and President of the Royal Academy of Engineering
Inspiration: a physics teacher who brought hairdryers and plugs into the classroom
Passion: flight
Mission: to make transport, especially aviation, environmentally friendly
Favourite invention: conceptual designs for a silent aeroplane
Advice to young engineers: 'Great opportunities will arise, grab them'
Date of broadcast: 26 August 2012

Professor Dame Ann Dowling designed a near-silent aeroplane. Determined to show the world just how quiet passenger planes could be, she led the UK side of the Silent Aircraft Initiative (a five-year collaboration between Cambridge University and MIT) and proved that there was no technical reason why an airport should be any more noisy than a busy street. A mathematician who was pulled into aeronautical engineering during an undergraduate summer job at RAE Farnborough, she wrote her PhD on noise-reduction strategies for Concorde. In 1993 she became the first female professor of engineering at Cambridge University and became head of the department in 2009, in charge of 150 academic staff and with a turnover of £74 million.

'Flight always has seemed quite incredible to me,' Ann told Jim. 'And I think perhaps it still does.' Even for an engineer who has spent most of her life studying the mechanics of flight, there remains something strikingly counterintuitive about seeing such hefty structures climbing into the sky.

'Did you imagine, when you were a child, that you'd end up spending your life thinking about working with engines?' Jim asked.

'I don't know what I imagined as a child, really,' Ann said, laughing. 'I think I would be very surprised by what I'm doing now. Absolutely astounded! But I was interested in aircraft and in flight.' In her early twenties she seized the opportunity to fly a plane herself. During a mid-conference outing in a seaplane over the mountains of Seattle, Ann found herself sitting in the co-pilot's seat and, having never flown a plane before, was invited by the pilot to take control of the aircraft and to 'see what it can do'. 'I looked back, and a colleague looked positively green!' she laughed. 'I think he probably knew how close I was to stalling.' With a bit of careful advice from the actual pilot, she then successfully brought the plane in to land without incident. She later trained for a pilot's licence and has been an enthusiastic private pilot ever since.

Much of Ann's professional life has also been focused on flight. She spent around a decade, on and off, designing an aeroplane that would be both significantly quieter and more fuel efficient than existing aircraft. But her career began a long way away from the noisy world of aeroplanes and engines, in the abstract world of mathematics. She graduated with a first-class honours degree in mathematics from the University of Cambridge. There was no doubting her mathematical ability, according to her friend and colleague John Moran, an aeronautical engineer for Rolls-Royce. 'I've never seen anyone get through so many pages of mathematical calculations so fast,' he said. He remembers one occasion where they were both involved in a minor road-traffic accident in Germany. To John's surprise, Ann's first reaction to the incident was to whip out her PC and show John her latest calculations. 'She lives for describing the world that she lives in in terms of mathematics,' he said. 'It's part and parcel of Ann; it's part and parcel of her life.'

'Do you enjoy the actual process of solving the equations themselves?' Jim asked.

'When they're going well, it can be very pleasant!' Ann said. 'Particularly when things come out that you weren't expecting.'

Solutions to equations can reveal surprising properties of systems, shedding new light on practical problems. The logical beauty of pure mathematics is enough to keep many a mathematician within the realm of the abstract, but Ann was always more interested in how mathematics could be applied to the world than in 'maths for its own sake'.

A summer spent working with Ted Broadbent at the Royal Aircraft Establishment in Farnborough introduced Ann to all the different noises made by engines. Suitably inspired, she decided to study noise-reduction strategies for Concorde, and progressed from a mathematics degree to a PhD in engineering. Test flights were under way for the world's first supersonic plane – a sleek, futuristic machine with an impressively sharp beak – that promised to cut transatlantic flight times in half, reducing the travel time from London to New York to three and a half hours. But many citizens stateside were concerned that the noise would be intolerable. Not unreasonably, it turned out. A paper released in 2004 showed that the noise made by Concorde during its first four months of commercial flights in 1976 regularly exceeded the average pain threshold for humans – that is, 110 perceived-noise decibels.

Following a wave of citizen protests in New York and Washington DC, the US Congress had banned such a noisy plane from landing on American soil and, in so doing, had thrown a major spanner in the works of this Anglo-French initiative. Without the coveted transatlantic routes, Concorde would definitely struggle to balance its books. Not a lot of thought had been given to aeroplane noise before; but there was a surge of interest in aeroacoustics, following the ban. 'There was this whole new field opening up,' Ann said. 'We just didn't know what to do – you know, if you change something, do you make it better or worse? . . . It was a really exciting area.'

Solving abstract mathematical problems was fun, but working out how to understand the real world using mathematical logic was even more enticing, for Ann. 'That's when things start to get messy,' she said, smiling. 'The real world is hugely complicated . . . If you take something like an aeroengine, it's got so many bits and it can behave in so many different ways, and they interact. You couldn't possibly solve that all at once. First of all, you need to decide what is most important and you need to try and model those elements; then you need to find some set of equations that can be solved that captures

> 'The real creativity is in correctly modelling the system in the first place. If the model is not correct, no matter how clever you are, the maths you're solving is just nonsense'

their behaviour – that's really where the creativity is, I believe. After that it's technique; once you've got a set of equations, it's like solving a puzzle. That can be quite satisfying. But the real creativity is in correctly modelling the system in the first place. If [the model is] not correct, no matter how clever you are, the maths you're solving is just nonsense.'

Ann created mathematical models that described some of the noises made by Concorde. By the end of her PhD she understood the noise 'much better' but had not managed to find a way to reduce it. Too many design decisions had already been made that severely limited her options and, after experimenting with all sorts of possibilities, she concluded that the noise problem could not be solved by add-ons. Meantime, the first commercial supersonic flight had taken off, flying from London to Paris in 1976, and the US ban was lifted the following year. The perceived need for business executives to save time trumped residents' desires to avoid being exposed to excessive noise.

While Concorde roared through the sky (1976–2003), noise reduction was not a priority in the aerospace industry. There was little funding for engineers who were interested in quieter planes. And Ann applied her knowledge of the complex mechanics of engine noise to studying submarines, cars, combustion systems, gas power generation and wind turbines. During a sabbatical at MIT, however, she spotted an opportunity. Cambridge and MIT had launched a joint initiative, keen to work with each other and with industry, to engage in some blue-skies thinking that could change our world, and they were looking for ideas. Picking up on her work on aeroacoustics, decades after her PhD, Ann and colleagues proposed a research project to explore the limits of noise reduction in aeroplanes. Just how quiet *could* they be? Their proposal was turned down the first time, but they rewrote it, submitted it again and succeeded. In 2001 the Silent Aircraft Initiative, a five-year research collaboration between Cambridge University and MIT, was launched. Concorde was designed and built before noise became an issue: countless engineering decisions had been made and there was only so much that could be done, in the final design stages, to make it quieter. The Silent Aircraft Initiative was an opportunity to start from scratch: to think about aircraft design, prioritising noise reduction together with fuel efficiency.

Determined to find out what was possible in terms of noise reduction, Ann assembled an international team of graduate students, professors and industry engineers (including her husband Tom Hynes, also an engineering academic at Cambridge University) to design a plane whose noise 'would be imperceptible beyond the perimeter of an urban airport'.

It was an ambitious target. To achieve it, the noise output from a standard passenger plane would have to be reduced to 0.3 per cent of the existing levels.

'That is a very dramatic reduction,' Jim said. 'What gave you the courage to believe that that sort of reduction was even achievable?'

'Well, I suppose we didn't know that it would be,' Ann laughed.

'Where do you even start?' Jim asked.

'Actually, with that it was very clear where we had to start,' Ann replied matter-of-factly. 'We had to start with reducing the jet noise. It's as simple as that. And we know that the only real way to get that kind of reduction in jet noise is to make the propulsive jet slower and bigger.' Slower jets are less energetic and the sound waves generated are much less intense. Since the amount of thrust you need to get the aircraft off the ground remains the same, the slower jets needed to be bigger to produce the same thrust.

Working backwards, Ann calculated just how big these slower jets of air exiting the engines would have to be to generate enough thrust to get an aeroplane off the ground. Then she asked herself what the aircraft would have to look like to generate and accommodate these larger jets. 'When we first did that, we just scared ourselves silly,' she said.

But if noise control was non-negotiable, they had to try to find a way. Faced with an apparently insurmountable problem, the team took a step back. Thinking about how to accommodate such large engines without creating excessive drag led to a radical redesign of the shape of the plane. Typically, aeroplane designers start with an outline of the shape: 'usually it's the tube and some wings, and "The engines will hang here."' Then they start to talk about minor changes to it, according to particular specifications. But if noise reduction was a priority, a 'blended wing–body' made more sense than the traditional 'tube and wings' passenger jets we're accustomed to flying in. It was a design that had been around for decades[1] but was

---

1 Wind-tunnel tests in the 1920s suggested that this would be a promising shape for planes, but the first plane to be built like this stalled on its first flight, severely injuring the pilot, and the British Air Ministry lost interest.

more commonly found in long-range bomber planes, not passenger jets. 'Some people have described it as looking more like a bat,' Ann said, laughing. 'It's much wider' than a standard passenger jet and 'there are many more people abreast.' As a result, there are far fewer window seats. 'First-class seats would have to be down the middle rather than up the front, because they will be less affected when the aircraft tilts to turn.'

In traditional passenger aircraft, only the wings generate lift. In the bat-like blended wing–body planes, the wide flat fuselage helps out. This increased lift means a better lift-to-drag ratio than cylindrical-bodied planes, which improves fuel efficiency. The increased internal volume provides a way of embedding the large engines (needed for low jet noise) within the airframe and so reduces the drag on them. And having the engine intakes on the top of the plane, ingeniously, was 'great for the noise as well'. The plane itself acts as a shield, blocking sound waves generated by the engines from heading down to the ground. 'Any noise from the engine that tries to travel forward and down bounces off,' Ann explained. And in this way a significant amount of the engine noise, produced on top of the airframe, can be deflected away from airports and surrounding residential areas into the sky above the plane.

'And what about noise inside the plane?' Jim asked.

'Terrible, I should think,' Ann laughed. 'We thought that the people who were choosing to fly had made their choice, whereas those on the ground weren't choosing to hear the aircraft.' The noise inside the cabin was not a priority for Ann, although 'I think you can do things about that,' she said. Sound absorbent wall linings or noise cancellation devices, for example.

According to Ann's calculations, radically changing the

shape of the plane would dramatically reduce the amount of engine noise that would reach the ground when the plane was airborne. But the sheer amount of noise generated by the airframe still needed to be addressed if there was going to be any hope of designing a near-silent plane.

'When a large passenger jet comes in to land, you get as much noise from the airframe as you do from the engines,' Ann explained.

An aircraft preparing to land needs to descend and to slow down. 'To do that, it's got to lose both potential and kinetic energy.' All that energy has to go somewhere. And since the total energy in a system remains constant, according to the laws of conservation of energy, 'the only way the aircraft [can do] that is to dump energy into the surrounding air.'

'Now you're talking my language!' Jim laughed. 'Now you're talking physics and the conservation of energy.'

A typical passenger aircraft approaching a runway loses about 20 megawatts as it comes in to land. Only the tiniest proportion of that energy escapes as sound (most of it escapes as kinetic or heat energy), but it's enough to make a lot of noise and be heard by our sensitive ears.

'The engines are almost the easy part,' Ann joked. 'At least you know where the noise is and where it's coming from!' Airframe noise (that is, the noise generated by airflows around the body of the aircraft) is notoriously hard to model.

Engineering isn't just about focusing in on one element of a design and absolutely perfecting it, Ann explained. You have to pull it all together, 'and that's what's difficult.' An engine designed to be as quiet as possible will impact on the plane's performance. This in turn might create more noise from another source – airframe noise, for example. Because everything is intimately interconnected, being a technical expert in one field (engine noise, for example) only ever gets you so far. It's the ability to pull all the parts together into an integrated system that makes a really great engineer.

Approaching the problem from many different angles and thinking about how the different parts of the system would work together, Ann and the team concluded that it was possible to make an aeroplane that would be, on average, 25 decibels quieter than the planes currently soaring through our skies. Put another way, the silent aircraft would generate less than 1 per cent of the sound energy created by planes today. And it would be environmentally friendly in another way: using 25 per cent less fuel per passenger mile than the most efficient planes in the skies today.

The Silent Aircraft Initiative was set up to find out what was possible in terms of noise reduction. It proved that if aeroplanes were to be designed in a different way they could be radically quieter than the deafening vehicles that are in operation today. There are no current plans for the conceptual design SAX-40 to go into production, but it has been designed and tested using industry tools. Rolls-Royce made their engine-design suite available to the team, the gearbox was checked by NASA engineers and Boeing provided the computer software to validate the blended wing–body design. Some aspects of the design have even been tested in wind tunnels. In short, Ann and the team have shown that it should be possible to design an aeroplane that makes barely any noise. The main barrier to such a plane being built is the enormous expense. 'We really never had any illusions about just how costly building such an aircraft would be,' Ann admitted.

Innovation is expensive. 'Any study of the aeroplane business says aeroplane manufacturers don't make money by bringing in radically different aircraft designs,' Ann said. 'The real money is made providing the same types of designs over and over again. Scaling them up and scaling them down. And in maintaining and selling replacement parts for current aircraft.' To produce just one prototype of the silent aircraft would cost an estimated £20 billion. And the expense doesn't stop there. Ever since the Boeing 707, commercial aeroplanes have stuck to the same basic rules, making mass production of fuselages and replacement parts as straightforward and cheap as possible. Introducing a radically new shape to planes would massively increase ongoing manufacturing costs, in addition to the upfront research and development expenses. Passengers would need to pay an awful lot more for their seats so that the residents living beneath their flight path could enjoy a quieter life.

'How many precious years of your life and the lives of your colleagues have been devoted to tackling this problem?' Jim asked.

'Oh, my goodness! For me, it's probably been ten years, on and off – doing other things as well – but it's been that sort of duration.'

'It must have been tremendously frustrating for you,' Jim said, trying to imagine what it would be like to spend a decade designing a plane, and then not being able to see it go into production. But, as Ann reminded Jim, the Silent Aircraft Initiative was always only meant to be a conceptual design. And that her team of researchers wouldn't have been the right people to plan the manufacturing and detailed design. The important thing, for Ann, is that they demonstrated such an aircraft was credible. They set a target for noise reduction that far exceeded what anyone had previously imagined was possible. Now no one can dismiss the idea of a near-silent aircraft as an impossible dream.

'We [didn't want] people to be complacent about what could be done in terms of low noise,' Ann said. 'I think what we wanted to do here was to change minds, and I think that we did actually achieve that.'

'We didn't want people to be complacent about what could be done in terms of low noise'

You don't need a real aeroplane to get real results. NASA performed wind-tunnel tests on a scaled model of SAX-40 to measure aspects of the aerodynamic design and noise reduction, confirming the team's predictions. And, as a result of those tests, have changed their noise targets. The double-decker, wide-bodied Airbus A380 is the quietest jet airliner on the market, and the next generation of Airbuses, the A320 fleet, will be even quieter.

The head of Boeing's Blended Wing Body Program (who was the Boeing contact on the Silent Aircraft Initiative), Bob Liebeck, said that a blended wing – body-based 'silent aircraft' could be built within seven years if funding was available. The question then becomes: who would fund such a radical new approach to flight? The commercial case for the SAX-40 doesn't add up. 'I think, if I'm frank, it will only happen if another type of funding comes along,' Ann said. Maybe a country that wants to break into the aerospace market would be interested. 'China is going into aviation in a big way . . . I think that's certainly a possibility . . . Or it might be because of a defence need. A quiet aircraft with a great space inside could be a great asset to the military. The 747 was precisely that: a military transport plane which then became a civil aircraft. And Boeing did quite well [out of it] for several decades.'

The Silent Aircraft Initiative proved that it is technically possible to dramatically improve the quality of life for millions of people who live near airports around the world. One day, perhaps silent aeroplanes will be standard. But, until the necessary investment is made, our skies will remain considerably noisier than our streets.

NAOMI CLIMER

*'My mission is to turn engineers
into rock stars'*

Grew up in: a village in Lincolnshire
Home life: married to another engineer
Job title: President of the Institution of Engineering and
Technology
Occupation: TV, radio and telecommunications engineer
Inspiration: building a radio station in Guernsey
Passion: making digital technology for the media
Mission: to show how creative and diverse a career in
engineering can be
Favourite invention: the tech that made it possible to
broadcast the BBC News Channel 24 hours a day
Advice to young engineers: 'Think beyond your discipline'
Date of broadcast: 16 February 2016

Naomi Climer is an IT engineer. As a girl, she wanted to play the cello for a living, but ended up inventing new technologies for radio, television, film and telecommunications. She 'fell into engineering' when she joined a BBC graduate training scheme for radio engineers, and worked her way up to become head of technology for BBC News. In 2000, she left the BBC to work on digital services for ITV. The tech she developed for ITV Digital (which didn't take off for commercial reasons) was later used in Freeview boxes. Working for Sony, she ran business-to-business internet services and as president of the Media Cloud Services in Los Angeles, she developed cloud-based technologies for Hollywood. In 2015 she returned to the UK and became the first female president of the Institution of Engineering and Technology, determined to change our perception of what it means to be an engineer.

'Why is it that we don't properly appreciate engineers [in the UK]?' Jim asked.

Naomi cut straight to the heart of what she thinks the problem is: the word 'engineer' seems to make many British people think of engines, not ingenuity. Americans might picture an Elon Musk-esque rock-star inventor living it up in Silicon Valley. In China engineering is considered to be the best possible training for almost every profession: politician, officer, chief executive or teacher. Germans typically treat engineers with the same respect as doctors. Meantime in the UK, 'there does seem to be a public impression of engineers as down in the boiler room, fixing stuff'.

'There does seem to be a public impression of engineers as down in the boiler room, fixing stuff'

Engineers serve the world in ways you might never have expected. 'We just seem to have the wrong stereotype for engineering,' Naomi told Jim. 'It's such a creative, incredible career . . . I can't believe that anyone would not want to be an engineer. It's baffling to me.'

Funnily enough, when she was younger Naomi herself actively did not want to become an engineer. She dreamt of being a cellist in an orchestra, but her father vetoed that idea. He said that anyone who was good at sciences and maths owed it to the world to do something with it. 'It seemed kind of pompous at the time,' Naomi recalled, 'but, as it happens, I wholeheartedly agree with him now.'

Having steered her away from her musical ambitions, Naomi's father then suggested that she might become an engineer. 'So, obviously, that was the last thing I would ever do!' she laughed. 'I went and did chemistry, much to his irritation, and loved it.'

After her degree Naomi thought, that's enough chemistry; what am I going to do now? 'I kind of fell into engineering,' she admitted 'not realising it was my vocation.' The BBC was running a graduate apprenticeship scheme for non-engineers with the right kind of aptitude, in order to get a more diverse group into their engineering staff. Naomi applied. Her future employers quizzed her, proactively searching for 'any hint of an interest in engineering', and eventually Naomi talked about how she once fixed her own car and mended the lawnmower.

Being a woman put her in the minority on the training course for sound engineers at BBC Broadcasting House in London, which had upsides and downsides. Some producers would see Naomi turn up to fix their radio studio on a night shift, and they'd think a prank was involved. 'This twenty-something-year-old blonde showing up – they'd think the boys in engineering were playing a bit of a joke! Little things like that happened but, to be honest, I loved working there.'

'So, it didn't bother you at all?' Jim asked.

'No, not at all! It was kind of entertaining . . . It did mean that you were under more scrutiny,' Naomi admitted, 'so every mistake that you made was absolutely amplified.' She remembers feeling humiliated when a tea break was turned into an opportunity to list all the mistakes she had made.

After a few years with the BBC, she moved to Guernsey to be with her husband. Looking for a challenge, she set up her own radio station on the island. Listening to the technology that she had built from scratch deliver its first broadcast gave her an enormous sense of satisfaction. It felt almost magical to have created something out of nothing in this way. 'That was great for me,' Naomi said, 'because instead of working in a big place where you had a very specific job . . . I was doing everything, from designing studios and building them and fixing them, to running operations.' Everybody pitched in to help out with whatever needed doing and, as well as using her technical expertise, Naomi went out on a motorcycle to file traffic reports for the breakfast news, stopping her bike every thirty minutes to call in live. 'There wasn't a whole load of traffic to report on, [so] it was more about the banter . . . I think that when you're in engineering you need to be versatile,' she said, smiling.

It's certainly true that people skills are – and will become even more so – a significant part of working in engineering. 'When you look forward,' she said, 'the kinds of challenges in engineering are so people related.' However clever our technical skill might be, the success of information technology depends as much on understanding what it is that people actually want or need. It's also, often, about getting people to work together, to work in teams to compromise on one or two of their ideals. In our ever more connected world, collaboration is inevitable and 'relationship building is going to be absolutely crucial to good engineering in the future', Naomi believes.

'Relationship building is going to be absolutely crucial to good engineering in the future'

Having followed her husband to Guernsey, Naomi then left the island when they decided to get divorced, and found a job back at the BBC. Before long she was promoted to become head of technology for BBC News. For this role, she needed to answer this question: how can engineering help broadcasters tell better stories? She brought in new kit to improve the quality of the pictures and videos we watch from our sofas every evening at six o'clock or ten o'clock. Live news is a 'very cutting-edge environment' and you're always on the move. Different stories need different technologies and different ways of connecting to the audience: 'One day you could be talking about how to armour a laptop to make it robust in the field, and the next you're thinking about how to move to widescreen or HD, or how to get things like mobile-phone contributions on the air.'

Asked about the next transformative technology, Naomi replied: 'I'm interested in three hundred and sixty-degree filming, where you film in all directions at once, because it allows you to tell the story in a much richer way.'

In some contexts 360-degree filming is already being made available. For example, with a rock concert it could allow the audience at home to take in more of the entertainment in a far more immediate way. But could there be such a thing as communicating news *too* vividly? There are editorial debates about what are the right things and the wrong things to cover in such detail. Would it be ethical, for example, to make a warzone visible in such an immersive way? How distressing would that be?

A number of broadcasters are developing 360-degree filming. Further questions are when and whether it will go mainstream, and whether it will be worth wearing the headgear. 'Not everyone wants to put a virtual-reality headset on,' Naomi said, 'but I think it has interesting possibilities.' If someone is bedridden, you could use the technology to 360-degree film their garden and give them an experience of being outside surrounded by flowers. 'There's a quality-of-life opportunity there.' Engineering doesn't just give us new ways of seeing the world, it can also create happier and healthier ways for us to live in it.

At the turn of the century, digital TV had a novelty on a par with today's virtual reality and in 2000 Naomi moved to ITV Digital to work with some brand-new technology. 'Pay TV is something we take for granted now, with the number of channels we have,' Naomi explained, 'but it was kind of radical at the time. There was a real "can do" feeling about ITV Digital. I absolutely loved it.' Freeview, the digital democratiser which brought multichannel television to anyone and everyone across the country, is built on the technology that Naomi and her colleagues worked on at ITV Digital. Their tech was well designed but it lost out to competitors for commercial reasons. (It didn't work so well for people who didn't have a good aerial.) Technological success relies on the gadgets that are created being on trend. 'You can have a fantastic technology that just doesn't fly,' Naomi said, acknowledging that success can 'come down to something as depressing as the IT not having been very well marketed.'

A couple of years later, Naomi started work at the Japanese multinational company Sony, where she would go on to become one of the most senior women in the business. She enjoyed the commercial pressure. 'I enjoyed being answerable to a number,' she said. 'It's absolutely clear whether you're delivering what the company needs you to deliver or not.' Despite the famed 'ruthless push' of big industry, Sony surprised her in being quite a values-based company. 'It didn't feel too dissimilar to the BBC,' she said. But she realised she had a lot to learn about Japanese etiquette. 'I learned it was very important where you sat at the table,' she said, 'and that it's important where you stand in the lift and who presses the buttons.' Formality was something she 'wasn't great' at. 'It's not really my thing . . . I've only learned since I left the company that it matters where you stand on the escalator, and I'm mortified at all the escalator gaffes I must have committed, not understanding that protocol!'

The Sony logo can be seen on speakers and TV screens. Back in 1979, the Sony Walkman introduced millions of people to the idea that they could listen to cassette tapes while they were on the move. (The Walkman even had a second earphone jack so that two people could listen at once, an idea that took several decades to catch on.) But Sony engineers do a lot more than create consumer products for us to buy.

Naomi ran Sony's non-consumer electronics in Europe, and worked on products ranging from security cameras to endoscopes and other medical equipment. 'There was a lot of system design and workflow,' said Naomi, 'thinking about the way things need to get from one end to the other.' She found herself filming operations in teaching hospitals and creating ways for students to watch and interact live with the surgeon. Afterwards, she needed to find a way to store that footage so that future students could always access it. It was a process similar to those used in the BBC newsroom: 'When you think that through, it's incredibly similar to running a live broadcast and then keeping it for repeats later.'

Away from the medical scene, she set a different plan in motion: using Sony's 4K digital projectors to make exclusive entertainment far more accessible. This made performances at the Royal Opera House and tennis at Wimbledon available live to people all over the world, without costing them a small fortune. At a live screening of the Wimbledon men's final in US cinemas, an American journalist was shocked to witness the cinema audience going deadly quiet when the on-screen umpire shushed the Wimbledon spectators on the screen. 'He described how totally immersed the audience had become. They were behaving absolutely as if they were in the crowd,' Naomi said, smiling. 'I do think you can deliver that kind of experience.'

She excels at making the intangible seem to come to life at our fingertips. But there's a certain technological intangibility that a lot of us struggle to visualise properly: 'the cloud'.

The cloud is 'on-demand computing'. It's about accessing things you might need (storage, computing power, or an application) from somewhere other than your own equipment. Google Docs or Spotify, for example. In 2012, Naomi moved to California and became president of Sony's Media Cloud Services, creating cloud-based technologies that allowed movie studios and production companies to share content and create productions without needing to be in the same room. 'You could have somebody in Australia talking to somebody in Los Angeles about a piece of video, frame by frame.' It can be very hard to explain which bit of a picture you are talking about when discussing technical issues on the phone and so Naomi and the team invented the video equivalent of highlighted text: 'mark-up tools that made it feasible to really talk in detail about something in a creative way.' It meant moviemakers could point to a particular bit of an image, while they chatted on the phone and explained what it was that they wanted to be changed. They could say, 'This bit of yellow is too yellow,' for example, while pointing to it on the screen. Or, 'Can you reduce the brightness here?'

Jim asked whether people were right to think of all their files and data floating up and being stored in 'this fluffy white thing' in the sky. Naomi said that wasn't a bad way of thinking about it. But the cloud does exist in geographical locations. That could be anywhere, from 'a server farm next door to your house' to 'a data centre near Las Vegas'. 'It could even be a building in Iceland,' Naomi explained. 'It will be physically somewhere, and you won't know where, and you don't actually need to know where. The important point is, you can access what you need.'

Theoretically, Naomi explained, there is no limit to the amount of data we can store in the cloud. 'I think it's made us all a bit lazy,' she said. 'We don't do as much housekeeping as we should because it's actually ridiculously cheap to just store everything for ever.' And so our hoarding instinct goes unchecked. People are more careful about tending to their own storage; they delete a few files if it's looking like they might run out of space. Storing information in the cloud sounds harmless enough. 'I just wonder if one day we will regret the amount of junk we've got in our cloud-based attic,' Naomi said.

'I just wonder if one day we will regret the amount of junk we've got in our cloud-based attic'

The cloud sounds nebulous. There is a definite feeling

of 'out of sight and out of mind'. But data centres are large, hot, concrete facilities that use staggering amounts of energy and generate thousands of tons of greenhouse emissions. The energy used by all the world's data centres in 2015 was considerably more than the entire electricity consumption for the UK. In 2018, data centres had the same carbon footprint as the airline industry, and demand continues to grow. One set of calculations predicts that the energy used by data centres will double every four years.

The cloud is an efficient storage system, but the benefits of this new improved storage technology have been dwarfed by the behaviour change they facilitate. (Washing machines are another good example of this. We invent machines that enable us to wash clothes and, as a result, we wash our clothes so much more often, that the energy saving that these machines could achieve is obliterated.) The existence of the cloud has allowed individuals (and companies) to be much less cautious about using up data-storage space. 'I think IT generally has a significant environmental impact and we don't talk about it much because we all rely on it so much,' Naomi explained. 'We almost don't want to know the answer.'

Given the devastating environmental impact of data centres, the association with white, fluffy clouds now seems alarmingly benign. In an attempt to reduce their reliance on non-renewable energy suppliers, many of the bigger companies are now choosing to locate data centres near renewable energy plants. Due to an abundance of hydroelectric power and ground-source heat-exchange generators, Norway is a popular spot.

}

Naomi chose to leave the bustling tech-hub of California and return to the UK in October 2015. She wanted to try and inspire the next generation of engineers, and had the opportunity to become the first female president of the Institution of Engineering and Technology. As our world becomes more and more interconnected with the internet of things, and our devices begin to talk to each other more and more often, it's important that we, as humans, keep talking about engineering and its necessity to modern living. 'I feel really strongly, especially with the internet of things and the cloud, that all of our lives are going to be so shaped by technology in the future and that we desperately need engineers to help us shape that world. All of the surveys that we've done say that we're not going to have enough engineers,' Naomi said.

'And does the public see how engineering is behind so much of our modern world?' Jim wondered.

'I think the answer is no, not necessarily,' Naomi admitted. 'Whether you're looking at the extraordinary mobile phone and thinking "wow" about the engineering in that, or a bridge,' she continued, 'or even a paving slab, or my clothes, there's engineering in absolutely everything. I would like people to just be a bit more conscious of that, to be thinking about the engineering behind things. And that just requires a lot more of us to be talking about it, more of the time.' We don't just need to talk about engineering more, we also need to start to talk about it in a very different way. 'There's fairly substantial research to suggest that part of the problem is that when

> 'There's engineering in absolutely everything. I would like people to just be a bit more conscious of that'

[women] look at the options in engineering, it just doesn't look like it's for them. They don't feel like they would fit in ... There's evidence that offices [with] *Star Wars* posters all over the walls and particular magazines lying around make many technically minded women feel, "Er, I'm not sure. This doesn't feel like me."'

When Naomi worked for Sony, their diversity 'wasn't brilliant'. 'I was routinely the only woman in the room, [which] can feel a bit lonely.' It is a situation that she is keen to change. 'A disproportionate number of graduate women drop out and go do something different,' she said.

'What about the argument that boys like things, objects, machines more than girls do?' Jim wondered.

'There's good research that says, on a statistical basis, "Boys like things, girls like people",' Naomi agreed. But, she believes, it's not a reason for women to be less interested in engineering than men. Engineering is about people shaping the world according to our desires. It is as much about people as it is about things. 'I do think companies could do an awful lot [more to attract women] just [by] thinking about the images and language they use to market themselves and recruit.' If only they could 'articulate the creativity, the humanity, the ability to make a difference'.

What if schoolchildren were to be taught about the real breadth and creativity of engineering as a subject? 'We need teachers to better understand what engineering is,' she said, 'from aerospace to healthcare to biomedicine, you name it! Teachers are enormously influential.' They can open young people's eyes to all the different things engineers do, and remind us all that engineering is about making the world the way we want it.

Naomi dreams of a time when engineers are treated like

rock stars. Perhaps because Hollywood is so close to Silicon Valley or perhaps because the information technology that has been created has generated such vast fortunes, Elon Musk-style entrepreneurs have celebrity status in the US. Maybe we are witnessing the beginnings of this in the UK: Sir Tim Berners-Lee, inventor of the World Wide Web, was greeted with cheers when he appeared at the opening ceremony of the Olympic Games in London in 2012.

Engineers might not be household names right now, but if people understand better the sheer creativity, the humanity and the ability to change the world that you might have as an engineer, Naomi will be happy enough. 'I would hope that by the end of my year [as president of the IET], people are more familiar with the idea that engineering is a diverse and extraordinary option.' Everyone should consider it.

TONY RYAN

'Specialist knowledge means you can go off in many, many different directions'

Grew up in: Leeds
Home life: married to Angela, who works in HR, with two daughters, Gemma and Maria
Occupation: chemist
Job title: Professor of Physical Chemistry at Sheffield University
Inspiration: the nylon rope trick
Passion: polymers
Mission: making sure useful materials are available to future generations
Favourite invention: catalytic clothing
Advice to young inventors: 'Beware of maintaining the status quo'
Date of broadcast: 21 February 2012

Professor Tony Ryan is a polymer chemist who finds nano-solutions to global problems. He started his career studying polyurethane and went on to collaborate with people working in a wide range of different fields, from fashion designers to tissue engineers. Distressed by the poor air quality in his home town, he invented catalytic clothing so that residents could clean up air pollution as they walked about town. In 2005, he was awarded an OBE for services to science, and in 2008 he became pro-vice-chancellor for the Faculty of Science at Sheffield University.

Jim Al-Khalili interviewed Tony a year after he had launched Project Sunshine, a multidisciplinary research project to find ways to harness more of the energy from the Sun and use it to meet our growing energy needs here on Earth.

'I came to polymers very early on,' Tony told Jim.

He witnessed his chemistry teacher performing the nylon rope trick and he was entranced – watching, amazed, as a stretchy thread emerged from a mixture of two liquids. As far as Tony, aged fourteen, was concerned, his teacher might as well have been pulling a rabbit out of a hat. The thread grew and grew, as his teacher pulled a paper clip out of a beaker of liquid, raised it above his head and trailed it backwards and forwards across the front of the classroom. There was, it seemed, no end to this magically manufactured piece of nylon string.

Tony explained how it worked: 'There's one chemical dissolved in the water, and there's another chemical dissolved in the oil.' At the interface between these two liquids, these two chemicals join together to form long chains 'made up of many little things'.

Chemists working for DuPont discovered the nylon rope trick in 1935 and then proved to the world just how useful this stretchy and strong synthetic polymer could be. Four million pairs of nylons were sold in four days when synthesised silk stockings first went on sale in 1939. And polymers (long thin molecules, made up of multiple repeating units) have been custom-made to meet our manufacturing needs ever since. In the case of nylon, which is a polyamide, the repeating units are molecules called amides. Other polymers are made with different linking molecules. Polythene (linked by molecules of ethene) proved useful for making bags. Polystyrene for packaging. Polyvinyl chloride (PVC) for all sorts of things. Polyesters could replace cotton. Cross-linked resins reinforced with glass made fibreglass. When Formula One cars, jumbo jets and rockets needed materials that were both lightweight and strong, polymers reinforced with carbon did the job.

There are plenty of naturally occurring polymers too, Tony reminded Jim. Jim wondered whether Tony looks at the world and sees polymers everywhere he turns. 'Well, I look at you and I see Jim,' Tony laughed. 'But I can look at you through my polymer-chemistry eyes and see a Jim that's skin and bone and muscle. Skin is a polymer and muscle is a different polymer.' Tony can see polymers everywhere on Jim – or nearly everywhere.

'Unlike you, I'm lacking any polymers on top of my head,' Jim said, laughing.

'I wasn't going to mention that!' Tony replied kindly.

'Your cotton shirt is made of polysaccharide,' Tony said, continuing to play spot the polymer. 'And I'm just going to peek at your trousers,' he said, with a smile. 'Jeans are made from cotton and the soles of your shoes are made from polyurethane. So, yes, I can see polymers everywhere.'

> *'I can see polymers everywhere'*

Inspired by the nylon rope trick and the teacher who showed him how to do it, Tony decided to study polymer science and technology at the University of Manchester Institute of Science and Technology (UMIST). He stayed on to do a PhD, specialising in polyurethanes, and was approached by Sheila MacNeil, a tissue engineer, who was trying to create an artificial cornea. The transparent layer at the front of the eye is easily scarred, and damaged corneas are a leading cause of blindness worldwide. Sheila needed a material that could be used to build a scaffold on which new corneal cells could grow. It needed to be porous so that cells could attach themselves to it, and highly flexible so that intricate structures could be created. Polyurethane foams and fibres are more commonly used in upholstery, but they had many of the properties that Sheila was looking for and she thought Tony might be able to manufacture a transparent polymer that would do the job.

Keen to deliver, Tony aimed high. He wanted to create the Rolls-Royce of artificial scaffolds: an intricate structure to match the sophisticated structures that our bodies create. 'But sadly, the cells didn't appreciate it at all!' Sheila said. On his second attempt Tony learnt that aiming for something 'more like a Morris Minor' was the way to succeed. Sheila, meantime, learnt that Tony is 'at his most creative when things go wrong'.

'For skin and the cornea you don't need a clever scaffold,' Tony explained. 'The scaffold can be dumb because the cells do all the clever stuff.'

'The scaffold is just the material that the cells grow on?' Jim said.

In short, Tony was trying to do too much. Once he focused on the scaffold and 'let the biology take care of the biology', he created a material that could convince our bodies that it was the real thing. The material he invented was tested in clinical trials in India and has since been used to help repair the vision of many people with damaged corneas.

For many polymer chemists, understanding the structure and properties of different polymers is an end in itself. Tiny changes to the chemistry of a molecule can have a dramatic impact on its properties, and it can take a lifetime of research to understand how and why they work. But Tony wanted his polymer creations to be useful. Collaborating with Sheila had shown him that a polymer intended for one purpose could perhaps be used 'in something else'. Specialist knowledge of how particular polymers form enabled him to 'go off in many, many different directions', Tony said. 'And I loved that.'

When the fashion designer Helen Storey called him up, he 'was really, really pleased'. She had heard Tony and Sheila talking about their artificial corneas on the Radio 4 programme *Material World*.[1] The combination of Tony's easy-going demeanour and scientific expertise struck a chord. She loved his curiosity and he sounded as if he would be approachable. The chemistry department in Sheffield didn't get many calls from fashion designers in London, and Tony 'was quite flattered that she'd picked up the phone!'

Helen asked Tony: 'If quantum mechanics tells us that a particle can be in two places at once, why can't packaging know when it is empty and disappear?' – and he had to stop himself from laughing out loud. They occupied very different worlds, but at the interface between their disciplines, they created something new: the world's first disappearing dress. It was made from a water-soluble plastic fabric and was designed to get consumers to think about the limited resources that are available to us all here on Earth. They hung these exquisite plastic dresses in the Meadowhall shopping centre in Sheffield for shoppers

---

[1] A weekly science programme on Radio 4 which has since been replaced by *Inside Science*.

to admire. When they saw such beautiful garments dissolving in front of their eyes, they were shocked. 'People would ask us, "Why are you destroying these beautiful things?"' Tony said. 'Which enabled us to ask them back: "What do you think happens when you go shopping?"'

The disappearing-dress project was called Wonderland, and the strapline was: 'Plastic is precious because it's buried sunshine'. The dress was a Trojan Horse, intended to raise awareness of our cavalier attitude to global resources. And it got Tony and Helen thinking about whether fabrics and fashion could be used to tackle environmental problems more directly. Cycling to work every day, Tony was regularly exposed to the noxious fumes from vehicle exhausts.

Every car is required by law to have a catalytic converter in the exhaust to help reduce some of the harmful emissions from car engines. The poisonous carbon monoxide, for example, reacts with oxygen in the air to form carbon dioxide, thanks to technology like this. Catalysts provide a surface on which particular reactions can take place and make them happen much more rapidly than they otherwise would. 'They take part in the reaction but are unchanged by it,' Tony explained. 'So, it stays there and can be used again and again.'

But catalytic convertors do little to reduce the nitrogen dioxide (a harmful gas that inflames the lining of our lungs) and is responsible for tens of thousands of premature deaths among city dwellers worldwide every year. Inspired by his foray into fashion, Tony wondered if he could create catalytic clothes that could help to clean up this polluting gas? Part of the problem with air pollution is that it tends to be worst in the places where most people live: there is more traffic in densely populated cities. Tony turned this problem on its head by making people part of the solution. What if city dwellers, dressed in catalytic clothes, could clean up polluted air while they walked around town, sat in roadside cafes or stood waiting for the bus?

During a 'really boring' meeting at the Royal Society of Chemistry, he started to calculate the surface area of his suit. If his catalytic clothing was going to have a chance of working, he knew the surface area in which reactions could take place would have to be as large as possible, and since he was wearing a suit, he took that as his starting point. Being a polymers man, he knew the diameter and the density of the PET fibres from which his suit was made. And he estimated its mass. Using this information, he was able to calculate the surface area per unit mass of his suit. From there, he moved on to jeans.

Before the meeting was over, he had concluded that if the combined surface area of all the clothes worn on a typical day in his hometown were to be converted to a catalytic surface, then it might make a real difference to the air quality in the city. 'If all the people in Sheffield wore catalytically enabled clothes, then they'd be able to take out enough nitrogen dioxide to keep us below the safe limit throughout the whole of the year,' he told Jim.

> 'If all the people in Sheffield wore catalytically enabled clothes, then they'd be able to take out enough nitrogen dioxide to keep us below the safe limit throughout the whole of the year'

Titanium dioxide is a known photo-catalyst (found in

sunscreen, paint and make-up) that can get rid of nitrogen dioxide. When sunlight hits molecules of titanium dioxide, the electrons that surround the nucleus in each of the atoms get excited. They get rid of their excitement by splitting the bond between the two oxygen atoms and the lone oxygen atoms that are liberated react with water to form hydrogen peroxide (H_2O_2). The peroxide, which is highly reactive, then reacts with 'whatever's around'. When it reacts with nitrogen dioxide, a harmful gas which inflames the lining of our lungs, it forms nitric acid, which, at suitably low concentrations, is harmless and can be washed away.[2]

And there was no need to buy a new suit. Any old clothes could be coated with titanium nanoparticles simply by putting them in a washing machine. Treating jeans with titanium dioxide would take denim, the original hard-working material, to another level.

Excited and keen to generate some enthusiasm for the idea, Tony and Helen 'set up a pop-up laundry so people could bring their clothes and have them done'. Catalytic clothing needed a collaborative effort to work. 'One or two people doing it won't have any effect . . . It's a whole-community thing.' Like herd immunity.

'You're a keen cyclist, aren't you?' Jim said, bringing the focus back to Tony. 'I have this vision of you dressed up in Lycra™ [another polymer] and, coated in this material, cycling around the streets of Sheffield, sucking up all the pollution like some sort of scientist superhero.'

Jim was joking, but Tony said that was precisely what

---

2 If nitric acid sounds alarming, bear in mind that both lemon juice and vinegar are acids. That said, no one would advise drinking nitric acid, however dilute.

he planned to do. 'As soon as I can treat the clothing of the guys I go cycling with, that's exactly what we'll do on a Saturday morning.'

Tony and Helen approached washing-powder manufacturers, hoping to inspire them to adopt this new technology. Ecover and Unilever were initially enthusiastic. They loved the idea of catalytic clothing, Tony said, but they were nervous about the negative publicity that might be generated by the use of nanotechnology. 'People are frightened of it . . . and they're frightened of it for all the wrong reasons.' There is nothing intrinsically dangerous about particles that are below a minimum size. Socks impregnated with nanofibres of silver to keep feet smelling sweet are popular Christmas treats. 'Soap was the original nanotechnology,' Tony said, and not many people are scared of soap. But in 2003, alarmed by a science-fiction novel by Eric Drexler called *Engines of Creation*, Prince Charles had raised concern that nanoparticles might self-replicate, and everything would disintegrate into a 'grey goo'.[3]

And there was another problem. The same mechanism that caused the catalyst to gobble up nitrogen dioxide also knocked out the molecules that were responsible for 'laundry fresh' aromas. If half the population of Sheffield wore catalytic jeans, the nitrogen dioxide in the air could be kept within the European Union recommended safe levels. The same is true for many UK cities.[4] But sadly,

---

3 Prince Charles later pointed out that he never referred explicitly to grey goo and retracted his comparison with the thalidomide scandal, while continuing to express his concern that more research on the social, environmental and ethical dimensions of nanotechnology was needed to ensure that it was used 'wisely and appropriately'.

4 For it to work in London, everyone would need to wear catalytic jeans.

Tony has not yet been able to persuade washing-detergent manufacturers to sacrifice the smell of their products for the health of our children and vulnerable elderly people.

Disheartened but not defeated, he arranged for a billboard to be erected on the side of one of Sheffield University's buildings that was made from catalytic fabric. It was able to remove the nitrogen dioxide generated by twenty cars every day and displayed a poem written by his friend Simon Armitage (who later became Poet Laureate) to raise awareness of air pollution.

Jim suggested to Tony that there are two different kinds of scientist: 'There are those who specialise in a very narrow field to become world experts. And then there are those, like you, who span lots of different areas. What makes you so convinced that your approach is the way forward?' Jim asked.

Tony laughed. 'Well, my approach is the way forward for me. If everyone was a big-picture person, then there'd be nothing to paint with. We need people to do the detail and to be very, very specialised, and to be absolutely the expert on that one thing. But we also need a community of scientists who make their expertise work for all of society.' For his commitment to this cause, Tony was awarded an OBE for 'services to science' in 2005.

He is committed to using science to find solutions to some of our biggest problems: food production and global energy supply, for example. 'As a chemist, you have to realise that the big problem we face has been caused by chemistry,' he said. 'Basically chemistry, in terms of making fertiliser, caused the problem because it allowed us to feed seven billion people.' Now he feels strongly that the scientists of the world are honour-bound, to come up with new ways of meeting our growing energy requirements.

Soon after he became pro-vice-chancellor of the Faculty of Science at the University of Sheffield, he set up Project Sunshine, a multidisciplinary research project focused on helping us to harness more of the immense power from the Sun. 'If we were able to capture all the energy that lands on the Earth from the Sun in one hour, we could drive the whole economy for a year,' he said. Capturing the entirety of the solar energy hitting each square foot of the Earth for any hour is an impossible dream. But it's a theoretical truth that shouldn't be ignored. 'There's so much energy available from the Sun, it would be stupid not to collect it.

> 'If we were able to capture all the energy that lands on the Earth from the Sun in one hour, we could drive the whole economy for a year'

'Until three hundred years ago, everything that happened on the Earth was powered by sunshine,' Tony said. He wants to create the technologies that will make this possible once again. 'The mathematicians work on how the Sun actually works,' he explained. 'The physicists work on converting the light that comes out of the Sun into electricity. The chemists make the molecules that capture the light and release the electrons. The same process happens in a leaf, and the microbiologists and biochemists study that. And the ecology group in botany works with the mathematicians.' So, you come around 'in a big circle with the Sun in the middle.' By bringing together scientists with very different interests and skills and setting them a challenge, he hopes Project Sunshine will help us to find a way to provide sustainable food and energy for the global population.

The most direct way to harvest the energy from the Sun is to capture sunlight in photovoltaic cells and convert it to electricity, and Project Sunshine has done a lot of work on new, cheaper, plastic solar cells. Previous attempts to make carbon-based solar panels aimed to make them as efficient as possible. Tony focused instead on keeping the costs down. He developed semi-conducting polymer blends that could be coated on a plastic sheet. His solar cells weren't as efficient as existing, traditional solar panels. 'They don't convert so much light into electricity.' But they were cheaper to make and could be produced on an industrial scale. They made access to solar power less of an impossible dream in sunshine-rich but financially poor parts of the world.[5]

'What motivates you more: is it the science, or is it that you want to improve the quality of life?' Jim asked.

'It's both, Jim,' Tony smiled. 'It has to be both. Why do we have scientists? We have scientists so that we know more about the world. Why do we want to learn more about the world? So that we can make the world a better place. They're kind of one and the same for me.'

---

5 The technology has been eclipsed now by perovskites (invented in 2012) and the ever decreasing cost of silicon solar cells, but when Jim interviewed Tony, plastic solar cells were on sale all over the world.

WENDY HALL

*'I was told there was no future for
me in computer science'*

Grew up in: London
Home life: married to plasma physicist, Dr Peter Chandler
Occupation: Web scientist
Job title: Professor of Computer Science at Southampton
  University
Inspiration: trying to put a documentary about Lord
  Mountbatten on a BBC Micro Model B
Passion: Web science
Mission: to help the world to understand the Web
Favourite invention: Microcosm, a multimedia informa-
  tion system that predated the World Wide Web
Advice to young scientists: 'Don't give up!'
Date of broadcast: 8 October 2013

Professor Dame Wendy Hall is a computing pioneer who hated computers when she first learnt to write code. She did a PhD in pure mathematics and taught mathematics for several years before joining the newly formed Computer Science Group at Southampton University in 1984, inspired by what computers could do for us. For many years she worked on creating a searchable database of the letters, audio recordings, photos and films of Lord Louis Mountbatten. In 1989, two decades before Google, and just as Tim Berners-Lee was inventing the World Wide Web, she invented Microcosm, a sophisticated hypermedia system (with links and search engines) that made it possible to access information from a wide range of different sources. She wants to create a more intelligent World Wide Web and has pioneered the study of Web science, an interdisciplinary endeavour to understand how the Web has evolved and help us to predict where it might go next.

Wendy Hall remembers being taught the coding language Fortran, as part of her mathematics degree, and being 'just so bored'. Every week she would hand in her work, a set of punch cards that she had prepared to instruct a mainframe computer, and every week it would be returned with comments from her teacher pointing out all her syntax errors.

'I was happy in the world of mathematics and really didn't see then that computers would offer me anything,' she told Jim.

'I was happy in the world of mathematics and really didn't see then that computers would offer me anything'

She didn't know what a PhD was, but the professors at Southampton University persuaded her to do one. 'I loved university, I loved Southampton, and I thought, another three years here would be great.' She had a wonderful supervisor and 'absolutely adored' her PhD, happier in n dimensions than in three. Ideally, she would have stayed in higher education but 'universities were on a down'. 'There were lots of cuts to higher education funding [in the late 1970s] and there were just no jobs. Certainly not in pure mathematics!' she said. 'So, I ended up teaching maths as a service course to engineers and then to trainee teachers.'

Five years later, the digital technology she disliked raised its ugly head again. Her bosses said: 'Wendy, you're a mathematician: you must be able to do computing.' She was asked if she would teach 'a bit of programming'.

Unenthused, she taught herself the Beginner's All-Purpose Symbolic Instruction Code, known as BASIC, during the summer holidays and started teaching a course in the autumn on the Commodore PET (Personal Electronic Transactor). This early PC could be programmed directly, which was a significant advance, but its display of lurid green text and occasional beeps did little for Wendy. It was the arrival of the BBC Micro Model B in 1982 that transformed her attitude to computing. It was hardly 'micro' by today's smartphone standards, but it was considerably smaller than the bulky mainframe computers that were in use in the early 1980s, mainly in university departments and offices. The BBC Micro looked like a glorified typewriter and could be connected up to a monitor or the TV screen at home.

'Suddenly it wasn't just about coding,' Wendy said. 'It was about putting pictures on to a screen!' The BBC Micro was designed for learning and for fun. It had high-resolution colour graphics and good-quality sound. (Too bad that the built-in speaker was prone to buzz.) It was a multimedia machine.

The other big advantage of the BBC Micro, for Wendy and hundreds of thousands of digital novices, was that there was no need to resort to machine code. The earliest personal computers could only understand numbers and were highly sensitive to syntax errors. The BBC Micro could process higher-level instructions more familiar to humans, such as:

```
10 PRINT 'I am the best'
20 GOTO 10
```

Wendy, however, spent most of her time creating graphics.

She was excited by the ways in which personal computers might help children to learn, and spent her spare time creating interactive lessons about electricity and anatomy. After a few successful sales, she set up a small company selling educational software that could be used on a BBC Micro. It went from strength to strength. And, a few years in, Wendy realised that she was spending a lot of time with her PC and less time with her first love, pure mathematics. When a job came up with the newly formed Computer Science Group at Southampton University, she 'took the plunge', joining the group in 1984. 'Then I saw these things called video discs,' she said. Video discs stored analogue information. Made of plastic, they were about the same size as an LP: the video equivalent of vinyl records, and some were called video LPs. They preceded digital versatile discs, or DVDs, which were invented in 1996.

The US entertainment giant MCA had put two hundred movies (including classics such as *Jaws*) on to DiscoVision laserdiscs (so named because they were read by a laser, not a stylus), but they were only ever enjoyed by a small-ish community of geeks. The quality was not much better than video tapes, which were widely available at the time and considerably cheaper.

Other uses were found for laserdiscs, however. Storing music and books required less memory than moving pictures. Audio CDs (compact discs) arrived in 1982. CD-ROMs (CDs with read-only memory) were invented in 1985. And in 1986, the BBC teamed up with Acorn Computers, Philips and Logica, and embarked on a pioneering multimedia project to mark the 900th anniversary of the Domesday Book. Written descriptions, photos, statistics and 'virtual walks' describing life in every part of the UK were stored on two laserdiscs, the Domesday Discs.

}

Inspired by the Domesday Discs, the archivist at Southampton University asked Wendy if she might be able to do
something similar for the extensive archive the university
had just inherited from Lord Mountbatten. Could this
vast collection of papers, photographs, LPs and films be
stored on a disc? Undaunted, Wendy set herself the task
of creating a system that would help the University of
Southampton to manage, store and retrieve all this valuable historical information. Most computer databases
did little more than manage payrolls or lists of addresses.
Wendy hoped to create a digital database that would link
letters, documents, photographs, sound recordings, TV
programmes and films. She worked out how to 'digitise'
letters, diary entries, telegrams and photographs (there
were no scanners in those days) and found a way to
convert analogue videos and sound recordings to digital
files. 'Putting video on to computer', as it was called for
many years, was a real challenge. As late as the 1990s,
multimedia was a very new idea.

Two decades before the arrival of Google, Wendy wanted
to create a system that would help people to access the
information they needed and suggest more material that
might be of interest to them: a photograph of an event or
person referred to in a document, for example. Or a sound
recording. To achieve this end, she created a searchable
database of 250,000 papers and 50,000 photographs,
films and sound recordings in which all the material that
related to a particular topic was tagged. Sadly, copyright
issues severely limited the material that could be made
public, and Wendy's plans for the Mountbatten archive
were curtailed.

'I was told once there was no future for me in computer science, or at the University of Southampton, if I continued doing what I was doing,' Wendy said. For most of her colleagues in the Computer Science Group, computing was about operating systems and computer coding languages, not information technology.

'*I was told once there was no future for me in computer science, or at the University of Southampton, if I continued doing what I was doing*'

'It sounds as if you were a bit of a maverick in an otherwise traditional computer science department,' Jim said.

'Oh, I was!' Wendy agreed. 'There were some profs who didn't think that what I was doing was computer science.' At that time, pursuing an interest in IT did not feel like the best career move.

A six-month sabbatical in the USA in 1989 changed everything. Wendy's husband Peter Chandler had got a great job offer in the USA and Wendy happily embraced a change of scene. The time she spent based at the University of Michigan in 1989 and attending conferences all over the USA 'was very, very formative,' she said.

She visited MIT and met the pioneers of hypertext, 'people like Doug Engelbart and Ted Nelson'. Doug invented the computer mous. He introduced Wendy to the idea that a computer user could interact directly with images on the screen and need not be limited to moving cursors up and down. Ted was the co-founder of the Itty Bitty Machine Company, and his thinking about personal computers had a big influence on the IBM PC. 'A user interface should be so simple that a beginner in an emergency can understand it within ten seconds,' he once said. Ted had started work on Project Xanadu in the 1960s on a mission to liberate words from the page. He developed a communications theory that transcended text and coined the term 'hypertext'.

Talking to Ted and Doug made ideas that Wendy had been grappling with when she was working on the Mountbatten archive come alive. She was not the only person in the world who was interested in linking things. With the help of a computer mouse and hypertext, she realised, 'we were also going to be able to interact by pointing at something, and that taking you to another document.'

The following year Wendy invented Microcosm, a multimedia information system. 'It was very advanced for its day,' Wendy said. The US pioneers had mastered hypertext, which was a way of linking documents. Wendy was interested in hypermedia, a way of linking images, sound and video files too. She was tackling problems associated with multimedia systems that still represent a challenge today.

In December 1990, Wendy gave a talk about Microcosm at the first European Hypertext Conference in Paris. Tim Berners-Lee was in the audience.

'We were concentrating on the idea that links were their own entities,' Wendy said. 'We stored them in databases and they had a meaning; there was a reason why a link existed. It linked one object to another. We weren't really worrying about the network.'

Tim approached Wendy after her talk and they started to collaborate. They were both very interested in the idea

of following a link to travel
from one document to another. Tim drafted a now famous
memo that went to his boss in
1989. The same year Wendy
ran the first pilot of Micro-

'There was a timeliness
about the need to share
information'

cosm. 'There was a timeliness about the need to share
information,' she said.

In Microcosm, links were stored in a separate database that needed to be managed. Tim's idea was to store
the links within the documents themselves and to connect disparate documents via the internet. 'Tim got the
network much more than I did,' Wendy said. 'He was
thinking about how to connect all the physicists working
at CERN.' Thousands of scientists working in universities
around the world were analysing the data produced by
experiments in the Large Hadron Collider and wanted to
share documents over the internet.

'The internet was there,' Wendy said. 'The research
community had started to use email. There were lots of
experiments with hypertext systems (such as Microcosm),
but they were all disconnected from the network.' Tim
brought these two ideas together.

'When you first heard about Tim's World Wide Web,
could you tell that it was going to work?' Jim asked.

'I don't think I really got it then,' Wendy said. 'I saw
him do a demo at the 1991 hypertext conference in Texas,
and I remember thinking how pretentious [it was] to call
it the World Wide Web, like the World Series in baseball.
But to him, it was going to be a worldwide hypertext
network – that's what was going to make it work. And
that's when I thought, "There's something very different
here."'

The World Wide Web was not an overnight success. Curiously, many of the very people who might have been expected to be the most enthusiastic were among the most sceptical. Wendy was keen, but plenty of hypertext researchers (who were working on their own linking systems) said that it would never work. 'I can remember reading papers by people working on hypertext that claimed to prove categorically that no one would use a hypertext system that had links in [it] that failed, or were dead or dangling.' How wrong they were! 'Now, along comes the Web. You get an "Error 404" and people just shrug their shoulders and say, "Oh well."'

For Tim's Web to be a success, it needed people to buy into the idea. Web browsers could only pick up documents that were written in HTML and, in the early days, not many such documents had been written. 'In the early days of the Web, there was hardly anything on it.' And when 'Error 404' popped up, as invariably it did (and still does when a link doesn't go anywhere), some early adopters had their prejudices confirmed. The World Wide Web was a flawed idea: writing HTML documents was a waste of time.

'Tim had to encourage people to put their information out as HTML documents.' He knew that without a critical mass of users, his dream of a World Wide Web would fail. And he insisted that the software would need to be open, free 'and, in places, broken'.

'Tim's legacy is that he gave [the standards and the protocols] away for free,' said Wendy. 'This meant that anyone, anywhere in the world, who wanted to make a Web application to access the internet would use the same standards and the same protocol. That's the amazing thing about it.'

Jim remembered discussing 'this thing called the World Wide Web' over coffee with his colleagues in his physics department at the University of Surrey in the early nineties. 'Some were sceptical that it would lead to anything much,' he said. Two decades later, it's hard to think of an area of our lives that hasn't been transformed by it.

'Has the Web developed as you thought it would?' Jim asked.

'I don't think I ever saw the scale and the speed at which it would change so many aspects of our lives,' Wendy admitted. 'I underestimated what people would want to do with it.'

She was, however, 'hugely inspired' by a 50-minute fantasy documentary film written by the science-fiction writer Douglas Adams, creator of *The Hitchhiker's Guide to the Galaxy*. *Hyperland* (broadcast on BBC Two in 1990) imagined how, in 2005, a steady supply of multimedia information would be delivered to us by a 'software agent'. Nine years ahead of Google, it anticipated how a search engine might work. In the film, Tom Baker (who had retired from his role as Dr Who) appears before Douglas Adams and announces: 'I am your software agent. I'm here to fetch and carry for you. I have the honour to provide instant access to every piece of information stored digitally anywhere in the world. Any picture or any film, any sound, any book, any statistic, any fact, any connection between anything you care to think of, you're only to tell me and it will be my humble duty to find it for you and present it to you for your interactive pleasure.'

'[*Hyperland*] does feel as though it's exactly what the Web is today,' Jim said.

Wendy agreed, but with some caveats. 'The things [Douglas Adams] talks about – the software agents, and the intelligent assistant to help us find the information – still doesn't really exist,' she said. 'That's the sort of system I wanted to build. [And] it's still being built.'

Wendy continued: 'The original Web was a web of documents and links . . . But part of the original vision was that it was a web of data, as well.' The big difference between documents and data is that people can read documents and computers can't. 'You and I can read a document and understand what it's about. We can decide whether a particular document will give us the answer to our question.' Machines can search for particular words, 'but they can't actually tell you what it's about', or if it's useful.

With data it's a different story. 'If you remember when you first used to look at train times on the Web, you just got a document and you had to read the document,' Wendy said. 'Well, now you interact with it.' This is made possible by vast amounts of cheap storage and very fast processors, and because the information has been made available as a set of data, not a document. A web of data, not just documents, is required if we are to enjoy the 'interactive pleasure' that Douglas Adams described in *Hyperland*.

'Machines can process and analyse data in ways that we can't possibly achieve, and as long as you tag that data to explain what it's about, a machine can then interpret the data to give you answers to questions . . . That is where we're going. It's what Tim called the semantic web, or the intelligent web. It's already there and it's beginning to make the Web an even richer information space. The semantic web is emerging,' Wendy said. 'And it's all about getting people to put data out. When the data's out, then you can start building the bigger ecosystem.'

'So is [the intelligent web] already taking off, would you say?' Jim asked.

'It has been a slow burn,' Wendy admitted. 'Nearly ten years ago now, I was at a web conference with Tim, and we were talking about why the semantic web, the web of data, wasn't emerging. Tim was very frustrated by it. And we started brainstorming about how the first Web had evolved, and we realised that, as technologists, we didn't have the answer. We couldn't explain it. We couldn't explain why some things had taken off and others hadn't.' Not being the types to accept defeat, they persisted and concluded that they needed to expand their horizons to succeed. 'You have to look at the technology map and you have to look at the social map, and the political map, and the economic map, and legal issues,' Wendy said. 'We realised that actually to study this ecosystem that we have created, we have to understand how the machines and people are working together collectively.'

Wendy and Tim and another British computing pioneer, Nigel Shadbolt, started to think more about Web-based technologies and the people who use them, and founded the Web Science Institute at Southampton University. 'The study of that is what we called Web science,' Wendy said. 'In some ways, the Web is an experiment that we

'In some ways, the Web is an experiment that we can't rerun'

269

can't re-run. It's changed everything about what we do and we're completely reliant on it . . . We take Google for granted now, but it arrived nearly ten years [after the World Wide Web], and it couldn't have been designed without the Web already existing.' More Web-based technologies followed: Wikipedia in 2001 and YouTube in 2005, for example. 'When Wikipedia started, people dismissed it as a silly idea that we'd never use.' It was full of inaccuracies and fairly empty when it launched. Not very *wiki* (the Hawaiian word for 'fast' or 'quick'). As more people used it, the quality of the service improved. 'But we could kill it as quickly as we have created it,' Wendy warned. The success of a technology depends on human behaviour as well as on silicon chips.

The master's course at the Web Science Institute takes students 'from really any discipline, from computer science, social science, law, economics, politics, mathematics, management, humanities. And they all end up as Web scientists. The computer scientists help the others learn how to write Web pages, and the social scientists explain human behaviour to computer scientists. They work on very interdisciplinary projects and they have to have two supervisors from different disciplines.'

The idea is to try to understand why some Web-based technologies have survived and thrived, while others died. 'Over the years, astronomers have trained their telescopes on the skies, taken those wonderful pictures, and then shared that data,' Wendy said. 'Our thesis is that, in order to study the Web, you've got to have a similar thing for the Web.'

To achieve this, she plans to set up a Web observatory to bring together as much information as possible from the digital world and help us to start to understand what on earth is going on. 'It's mind-bogglingly difficult to think about,' she said. 'There are researchers all around the world, in universities, in research labs and in industry, who are doing this type of work, but it's not joined up. And the companies that are doing most of the work – the Googles and the Facebooks –

'The Web observatory is about observing the Web'

own all the data themselves ... The Web observatory is about observing the Web. It's not about observing people on the Web, it's not Big Brother. It's about observing [what] we are creating [online] day after day after day.'

'We need the computer scientists,' Wendy said. 'The world needs more computer scientists, full stop. But this is different. It isn't all about coding. This is about studying this new kind of ecosystem, and I think our students are getting the most fantastic jobs because, increasingly, businesses and government need to understand how this whole [digital] ecosystem works.'

'The cool thing for me about Web science,' Wendy said, 'is that we get as many female students as male students; because they come in from such different backgrounds and because we're talking about how the Web is impacting on health, education and things that interest women so much.'

'Traditionally, computer science departments aren't overwhelmed with women,' Jim said.

Wendy sighed. 'I wrote my first paper on the lack of women in computing in 1987,' she said. 'And it's a hugely deep cultural thing that has developed in the West. It's

'I wrote my first paper on the lack of women in computing in 1987'

actually better in southern Europe than it is in northern Europe. America and northern Europe are about the same. And you sometimes think, well, is it a rarity that a girl likes computing? When I travel the world and I go to India and Malaysia and the Middle East – not so much in China, but in South-East Asia generally – I go into a computer science class in a university, and it's full of women. Computer science is a subject they want to study in those countries to get on, and the girls do as much as the boys. So it isn't deeply genetic . . .'

'It's nurture and not nature?' Jim suggested.

'It's very cultural,' Wendy replied. 'And we just don't seem to be able to turn it around . . . I want to get more women in. It really matters. As the lovely Karen Späck Jones [the British computer scientist who made search engines possible] used to say: "Computing is too important to be left to men."'

'All the evidence shows that girls do better at science and maths in an all-girls environment,' Wendy reminded Jim, then added: 'I think that's an important point.' When Wendy attended Ealing County Grammar School for Girls, she was able to enjoy her passion for mathematics 'without having to worry about boys'. Nevertheless, when she graduated with a degree in mathematics, she still imagined that she would become a teacher and get married. 'That's what most girls did in those days,' she told Jim. She married and she was a teacher for several years but, luckily for us all, societal norms didn't stop her thinking about how computers could be useful.

MARK LYTHGOE

'I am always up for the happen-stance event'

Grew up in: Manchester
Home life: dad of two boys
Occupation: medical imager
Job title: Director of the Centre for Advanced Biomedical Imaging at UCL
Inspiration: Isky Gordon, who gave him his big break into the world of science
Passion: mountain biking and medical imaging
Mission: to see things no one has seen before
Favourite invention: an MRI scanner that has the potential to destroy cancer cells
Advice to young inventors: 'Don't worry about the future, enjoy the present'
Date of broadcast: 2 September 2013

Professor Mark Lythgoe develops techniques for seeing inside our bodies in astonishing detail. And he has found a way of using MRI machines to steer therapeutic agents to precise locations in the body. At school, he found motorbikes and clubbing much more exciting than studying and he failed most of his A levels. Without having been to university at all, he managed to talk his way into doing a master's degree in behavioural biology. And after a few more twists and turns in his career, including working as a dog trainer in Israel and fitter in a factory in Manchester, he created one of the largest medical imaging research centres in the world at University College London.

'My roots are probably in disco,' Mark explained to Jim. If you ever swung by the northern music clubs in the early eighties, chances are that Mark could have been dancing on a podium in the same room. 'I absolutely loved music. I think dance music and the post-punk era and the Hacienda were all an outlet for a lot of my energy at that time.' When he wasn't dancing, Mark was spending his youth either scaling a mountain face or hurtling downhill on a mountain bike. School was never as captivating as riding motorbikes or the Manchester music scene.

'I just couldn't engage with the subjects that I was being taught,' Mark admitted. 'I didn't understand how they would affect me – my life, my friends, and the world around me. And I didn't find one single

'I remember my dad just crying with disappointment'

person that could bring these subjects to life for me.' At one stage he was excluded from school and ended up passing just one of his A levels, scraping an E grade in physics. 'I remember my dad just crying with disappointment,' Mark said. 'He was broken-hearted.'

After he failed to gain the qualifications he needed to make it to university, Mark's mum saw an advert in a newspaper for Salford College of Technology. Mark applied and completed his diploma in radiography, spending time at Manchester Royal Infirmary and learning about X-rays and ultrasound. But he had no real interest in the subject and wasn't offered a job as a radiographer after completing the course.

This first job was as a fitter at a plastic-moulding firm in Manchester, where he made plastic pipes. 'Around that time I met a chap who'd just come back from Israel,' remembered Mark. 'Next thing I know, I'm on the bus going to Tel Aviv and training to be a dog trainer, five kilometres south of the Lebanese border in the Golan Heights.' Mark described his job title as 'glorified pooper scooper'. It was not very intellectually rewarding, but outside of work he met people who had been to university and 'knew about stuff.'

'I wanted a slice of that pie,' Mark told Jim. 'I wanted something that they had.' Having been told that a PhD was the highest qualification you could get, he set his eyes on a new ultimate goal. 'I thought, that is it. One day I'm going to have a PhD. One day I'll make my dad proud of me.'

'So, this wasn't a case so much of you falling in love with science,' Jim asked, 'as a determination to prove to yourself that you weren't a failure?'

'Completely. I was completely driven by that ... I'd have counted baked beans for a PhD – just to be able to get to that level.'

Soaring up the academic ranks was never going to be easy given Mark's lack of qualifications, but he was highly motivated by a crushing sense of having failed. He came back to London and managed to get on to another diploma course at Middlesex Hospital in nuclear medicine. His determination to knuckle down saw him through a year there, but he knew he 'hadn't quite found' his vocation, yet. So he decided to spend some time working as a radiographer in Australia, and joined a project that was taking X-rays of people living in aboriginal settlements to work out the incidence of tuberculosis in the area. 'That was my first taste of research,' said Mark. 'I felt like I was doing something that no one else had done.'

'I felt like I was doing something that no one else had done'

At that point, he knew he had to somehow make the transition from radiographer to researcher: technician to academic. No mean task. It took a fellow non-conformist in the form of Professor Isky Gordon, radiologist at Great Ormond Street Children's Hospital, to get him there. Isky took Mark on as an assistant radiographer to help him with his research on imaging kidney diseases. 'Mark learned statistics to a level where he started teaching me!' Isky said.

'I was completely obsessed by it,' Mark recalled. 'I was finding out things myself while I was doing the reading, thinking, "Oh gosh, really, is that how it works?" And then you'd read a bit more and a bit more, and get to the point where actually nobody knows how that works. And I *love* trying to fill in the gaps.'

Isky helped Mark get on to a part-time MSc course and allowed him to take time off to complete it. To this day Mark remains deeply grateful to Isky for giving him his big break into the world of academic research, and also for helping him to be able to see things that no one else has seen before. This is the most beautiful thing about science, he believes. Perhaps for the same reason, he is a keen mountain climber. 'When you're climbing,' Mark explained, 'there is a very special moment when you stand on part of this Earth that nobody else has ever stood on, and you look out to a view that nobody else has ever seen.

That is one of the most special moments you can have in your life.'

'That feeling is identical to the day when someone gives me a call from the lab and says, "Mark, Mark, you've got to come down and see this!" And you see a picture from one of the scanners that nobody else has ever seen – a view of the world or the body that you just haven't imagined. You know that feeling – the hairs stand up on the back of your neck. That is what makes science so similar to climbing.'

'You see a view of the body that you just haven't imagined'

When he was working in Isky's lab, Mark tested a technique called autoradiography, 'which is when you inject a radioactive material into a person or an animal and then you can image that with a camera.' Mark and Isky had been trying to develop a new way of imaging the brain of someone who had suffered a stroke using radioactivity. But when Isky introduced Mark to David Gadian, his focus changed. David ran an MRI unit and offered Mark the opportunity to do a PhD and make as much use of David's high-tech MRI kit as he liked.

At first, David wasn't sure whether Mark had enough focus to complete his PhD. 'I think the main problem in the beginning, to be honest, was that he had so many other interests,' David said. 'But he has this remarkable capacity for taking on many things and doing them all at once without seeming to be too stressed by it all.' If Mark failed to turn up to the lab, his fellow researchers would picture him clinging on to a rock face somewhere or other. 'I think Mark is probably very good at figuring out the right path to take to get to the end point, whatever that may be,' David said.

}

Mark's path into the realm of science was unusual and convoluted, but once he had finally worked out where he was going, he was determined to get to the top. In 2005, he set up the Centre for Advanced Biomedical Imaging, or CABI as it is affectionately known, at University College London. It is now one of the leading biomedical imaging facilities in the world. And the secret of his success: a bit of everything. Harnessing his inability to focus on one thing at a time and turning it into a great strength, he created a centre where many very different things happened in the same place. Having noticed that the people concerned with biomedical imaging didn't talk to each other, he tried to bring them together. The ultrasound unit, the nuclear medicine unit and the MRI unit were all in different places doing their own thing. Mark was interested in all these ways of working, and so he was able to identify the crossover between these different techniques. 'I wanted to create a space that brought all the different disciplines together': mathematicians, quantum physicists, physiologists, chemists, anthropologists and artists. 'Everybody comes together.'

Having surgeons on the team was a useful reminder that the reason we want to identify tumours is to be able to treat them. And it led Mark and the team at CABI to come up with an ambitious plan. What if they could adapt MRI scanners so that they would search for tumours and then destroy them? In 2010 they discovered a way to move magnetic cells to tumours in the body using an MRI system. These 'magnetic seeds' could then be used, they hoped, to heat up and 'zap' cells in precise locations of the body that had been identified by imaging. And if everything went to plan, a new class of MRI scanners (which combine imaging with treatment) could be developed and used to provide highly targeted therapeutic interventions for prostate and lung cancer.[1]

}

In both his professional and personal life, Mark's energy is rarely contained to a singular outlet. Most imaging centres tend to be committed to one way of doing things but rather than focusing on one practice of imaging, or falling into the hackneyed battle between MRI and ultrasound, the CABI is home to twelve novel imaging techniques, 'all designed to help doctors to gain more insight into the inner workings of their patients' bodies and to better diagnose disease.'

'Which techniques are practised within the centre?' Jim asked.

'The classic magnetic resonance imaging – everyone will have heard of that. There's ultrasound, CAT scanning . . . But a new technique that we're working on is called molecular imaging. Most of the time, MRI and CAT scanning is just very good at anatomy – structure and anatomy. But we've developed a new technique; we've been able to sensitise the MRI scanner to pick up glucose.' Tumours take up glucose: 'They really eat it up!' The sensitised MRI scanner can pick up the sugars as they are taken up by the tumours.

---

1 Seven years later (and after this interview was recorded), Mark and the team showed that it was indeed possible to heat 'magnetic seeds' which had been guided through tissue directly into the tumour, and so destroy cancerous cells. They now hope to set up clinical trials to test the use of this new minimally invasive method for treating prostate cancer.

The tumours then light up on the scanner. All the patient needs to do before their scan is drink a bottle of Lucozade. 'Traditionally, you would have had to use a radioactive cocktail, attach it to the glucose and inject the glucose.' This new process makes the imaging available to everyone: because no radiation is involved, cancerous tumours can be detected in young children and pregnant women. It also gives a detailed view of molecular processes working on a tiny scale. 'With these new techniques,' Mark explained, 'we're hoping to be able to image cells or individual cellular processes. And that will transform things.'

The techniques that Mark develops give us an insight into parts of the body never seen before. 'We have a lovely technique that has been developed by Paul Beard's group at UCL, where they convert light into ultrasound inside the body,' Mark explained. 'It's called photoacoustic imaging.' Light doesn't travel through our bodies, but nanosecond pulses of infrared light can penetrate the skin. 'The [laser] light comes into the body and is absorbed by the oxygenated haemoglobin in blood vessels.' When this happens, the molecules in the blood expand, making a tiny sound (not unlike the way lightning generates thunder in a storm, but on a very much smaller scale). 'The sound waves generated by this barely detectable "crack" can then be picked up by the ultrasound detectors,' Mark said. (Light cannot escape the body but sound can.) 'And you end up with wonderful, three-dimensional pictures of blood vessels surrounding a tumour,' for example. The amount of detail in the images is 'extraordinary'.

Once they had established a reputation for being able to see the very, very small, Mark's team were asked to 'look for holes in the heart of a mouse embryo', to help geneticists who were looking for the gene that causes hole-in-the-heart babies. Using a super-strong 3D MRI scanner, Mark's lab created a technique sensitive enough to observe these tiny holes.

On another project, with Dr David Keays, Mark used state-of-the-art imaging techniques to visualise the cells inside the beak of a homing pigeon. In doing so he disproved the popular theory that migratory birds' navigational skills are the result of magnetic nerve cells in their beaks that enable them to detect the Earth's magnetic field.

As a young man Mark had many interests, and the projects undertaken at CABI are not always exclusively scientific in nature. He has collaborated with a number of artists over the years. There had been 'quite a lot' of ups and downs working with artists, but his favourite 'sci-art' project had a powerful effect on him and his belief in science.

British artist Andrew Kotting had the idea of collaborating on a project about his daughter, Eden, who has Joubert syndrome. Her cerebellum – the part of the brain where automatic processes, like walking, are stored – is missing. Every morning for Eden is like having to learn to walk all over again. Andrew wanted to take a look at the world through his daughter's eyes. He and Mark ended up working together for four and a half years.

'Very arrogantly, I thought if we just used science and knew everything about Eden's condition, I would understand how she looked at the world,' Mark admitted. Eden couldn't tell others how she saw the world and Mark 'thought science could provide that'. 'And I was very humbled by the project,' said Mark. He realised that

he never was going to get a
handle on the view that Eden
had of the world around her.
'Science cannot do the subjective experience, and it was

*'Science cannot do the
subjective experience'*

such a revelation. A hole in the scientific model. It was a
wonderful journey that I went on.'

'When I came into science I loved it and was obsessed
by it,' Mark said. 'But it changes you.' He said he learnt to
talk in a particular way and to write papers in a particular
way. He even started to walk and dress in a particular
way. The realm of science was incredibly exciting. 'But
I lost a little bit of Mark Lythgoe along the way,' he admitted. 'It was when I met Andrew and spent those four
and a half years working with him that he gave me the
confidence to be Mark Lythgoe again; to be an individual,
to have a personality and talk about my science not in a
completely objective fashion.' It gave him space to put 'a
little bit of Mark Lythgoe' back into the equation. 'It was
Andrew that gave me the confidence and the openness to
be able to do that. I do not believe that I'd be the scientist
I am today if it wasn't for Andrew. I can't quantify it.
Somewhere deep down inside, I feel a different person.
And because of that I think about science differently and
I do my science differently. The centre would not be the
centre it is today without those interactions that I've had.'

'I still cannot help myself being distracted, in a beautiful way, to go off and read something or see something or
get involved with something else that I love,' said Mark.
Trying to do everything has costs attached: 'It took me
longer to get the end result,' Mark admitted. 'What I hope
is that the end result was better for those digressions that
I have had.'

AILIE MACADAM

'Engineering can be really creative'

Grew up in: Chrishall, a small village near Saffron Walden
Family: married to Ade Sofolarin with two children, Yemisi
and Femi
Occupation: engineer
Job title: Managing Director of Bechtel Infrastructure
Inspiration: working on a sewage plant
Passion: urban infrastructure
Mission: to inspire the next generation of engineers, par-
ticularly women, to build their own future
Advice to young women engineers: 'Be authentic; resist
the temptation to conform to male-dominated norms'
Date of broadcast: 27 February 2018

Ailie MacAdam has worked on some of the biggest transport infrastructure projects in the world. She began her career doing experiments in a sewage treatment plant and was a chemical engineer for several years before she caught 'the construction bug' working on the 'Boston Big Dig', an ambitious civil engineering project to move a seven-lane highway underground. Seven years later she returned home to the UK to run the transformation of St Pancras station in London, turning an outdated station into an international terminal fit for Eurostar, while preserving its historic features. Next, she took on the Central London section of Crossrail, overseeing construction of 42 kilometres of new tunnels and seven new Underground stations under one of the busiest cities in the world. Thirty per cent of Ailie's top team of engineers were women. Not good enough, she says.

Ailie MacAdam's first job as an engineer was not the most glamorous. 'My first industrial placement was in Stuttgart at the Institut für Siedlungswasserbau Wassergüte und Abfallwirtschaft, which sounds a little bit better in German than it does in English,' Ailie said, breaking into a broad smile. 'It was a sewage treatment plant, which I loved.'

'It was a sewage treatment plant, which I loved'

'Dare I ask what that involved?' asked Jim.

'I'd put my white coat on and my wellington boots, gloves, safety glasses and hard hat, and I would go out and collect raw sewage from the front of the plant. Then I'd bring it back into the lab and do experiments.'

Set up to train students in the many different ways of treating sewage and to encourage research and development in the area, the ISWA in Stuttgart is the only sewage treatment works in the world that is attached to a university.[1] Typically, chemicals called flocculants are added to the sewage to encourage particles of sewage to stick together. The sludgy mixture is then poured into enormous concrete vats and stirred. Clumps of sewage settle to the bottom and the cleaner water is siphoned off.

Ailie's job was to determine the most efficient way of extracting the sewage that was suspended in the cloudy liquid waste. Adding different amounts of the flocculants, changing the speed of mixing, and varying the temperature

---

1 With 10 billion litres of sewage produced every day in England alone, it's a good job that someone cares about how it's processed.

all affected the rate at which the solids would settle, and by systematically changing each one of these variables and plotting the results of all these experiments on graphs, she was able to establish the optimal operating conditions at the plant. Based on her findings, the operators would then 'go and turn the dials on the big plant outside.'

'I've never heard anyone describe what we flush down the toilet so eloquently and with such enthusiasm,' said Jim.

'It was a fantastic experience,' Ailie replied. 'And you do feel a certain empowerment, actually, because I had to make sure what I'd done was right.' The results of Ailie's experiments determined the fate of sewage in Stuttgart. Her conclusions had consequences in the real world. 'So there's a sense of responsibility there as well.'

'Even though you're working with sewage, it sounds as if you felt respected?' Jim said, surprised that anyone who spent all day handling sewage could feel so good about themselves.

'Absolutely!' Ailie replied with enthusiasm. 'The contribution engineers make in all sorts of different industries is recognised in most mainland European countries.' Britain is the sorry exception in this regard. In Germany, Ailie was amazed to find herself, as an engineer, being placed 'on a par [status-wise] with medical doctors'.

The last of three six-month placements that Ailie completed when she was studying chemical engineering at Bradford University was with the multinational engineering company Bechtel. She joined their graduate training scheme directly after she graduated and spent ten years travelling the world, working on different projects in the oil and gas sector in Iraq (before the Gulf War) and at a paint recycling plant in Holland. 'What really made me jump out of bed in the morning was the environmental side of chemical engineering,' she

'I really got into understanding how bacteria can break down effluent'

said. 'I worked on environmental impact assessments in Romania and in an effluent plant up in Grangemouth . . . I really got into understanding how bacteria can break down effluent, which was terrific.'

The effluent from the oil and gas refinery in Grangemouth needs to be cleaned up before it can safely be discharged into the Firth of Forth. Part of the process involved adding different (aerobic and anaerobic) bacteria into the mix at different stages. Just as, under the right conditions, some microbes turn unwanted food into nutritious soil on a compost heap, so some other bacteria like to chomp on oil. Ailie's job was to determine the best operating conditions for these carbon-eating microbes and so improve the efficiency of the plant. What temperature, for example, would most encourage them to eat the oil-based pollutants that needed to be removed? Designing a treatment plant from first principles and bringing it into operation was satisfying work.

Working couples often have to make professional sacrifices: putting their own career on hold so that their partner can move on. But in Ailie's case, following her heart proved to be an excellent career move. Her husband, Ade Sofolarin, who is also a chemical engineer, was offered a job in Boston, and Ailie 'wanted to be there as well'. She asked Bechtel for a job in Boston and was moved into the infrastructure division to work on the Boston Central Artery Project. The Boston Big Dig, as it came to be known, was designed 'to build tunnels under downtown Boston and put a seven-lane highway underground.' To achieve this, 11 million cubic metres of earth needed to be moved. That's half a million truck loads. And if that's hard to visualise, try this: lined up bumper to bumper, the queue of trucks would span more than 4,000 miles.

'It was a hugely ambitious project, wasn't it?' Jim said. 'I remember people comparing it to the Panama Canal . . .'

Moving the artery underground involved building some of the deepest tunnels in the world, underneath downtown Boston and the harbour. The tunnel ventilation system alone needed seven buildings. Shutting a seven-lane highway and telling the cars to 'go somewhere else whilst you build and construct tunnels' wasn't an option. 'We had to keep seven lanes of highway traffic working and keep Boston running at the same time.' The old elevated artery was kept open while tunnels were excavated beneath it. 'It really was extraordinary. It is often described as being like performing heart surgery on someone as they're running a marathon.'

Before the project started, an elevated section of the seven-lane Interstate 93 Expressway was supported by a steel structure that ran between the city and the coastline, cutting across downtown Boston. 'They used to call it the green monstrosity,' Ailie said. The 'artery' was clogged with traffic most of the day. The air quality was terrible and the accident rate was four times higher than the

national average for urban interstate highways. 'Something
needed to change.' Analysts
were forecasting traffic jams
lasting 15 to 16 hours every
day by 2010. 'Now the traffic
goes underground and there's
beautiful parkland where

> 'Now the traffic goes underground and there's beautiful parkland where that monstrosity used to be'

that monstrosity used to be. People can walk from the
city to the coast. And people who live on the coast can
walk to the city. I'm very proud of that project,' Ailie
said.[2]

She had never imagined herself building roads, but
working on the Boston Big Dig, she was smitten: 'I'd only
been there a couple of weeks and I knew that this was
what I wanted to do with my life,' she said. 'I really got
bitten by the urban infrastructure bug.'

'Who hasn't been bitten by urban infrastructure at
some point in their life?' Jim said, smiling.

'I loved the challenges of delivering the infrastructure
in a way that kept all the stakeholders happy.' Working as
a chemical engineer was all about turning raw materials
into a product. Civil engineering, as Ailie first experienced
it, was all about people. Her first job on the Boston Big
Dig was to liaise with local residents to try and minimise
any adverse impacts on their lives during the construction
process. 'They all [had] their own needs in order to be
able to tolerate the disruption.'

---

2 The Boston Big Dig achieved many things, but it was not without
problems. It opened in 2004 (95 per cent complete), six years later than
planned. And, at $15 billion ($22 billion if you count the interest ac-
crued), it cost nearly seven times more than the original budget, according
to the *Boston Globe*.

In one area of downtown Boston, many of the residents worked night shifts and were complaining that the vibrations from the deep tunnel drilling were keeping them awake. So Ailie and her team first measured the vibrations to assess the nature and extent of the problem, and then worked with an isolation specialist to try and find a way of preventing the vibrations from disturbing these residents who needed to sleep during the day. The solution was neat: put the legs of their beds on springs, to deaden the vibrations. 'It's a good example of how lots of the engineering that goes into these projects is really creative,' Ailie said.

}

'You must have made quite an impact,' Jim said, 'because within a few months you had been put in charge of a $300 million project to build a new bridge . . .'

'Both our children were born in Boston, and I was fortunate to come off my first maternity leave to take over managing this project. It was bridges and ramps and some roads as well. It was fantastic.'

'Is it true that part of this multi-lane Boston highway is supported by blocks of expanded polystyrene?' Jim asked. 'The same stuff that we use in packaging?'

'Yes. We put these blocks together and pile them up on top of one another, to create a triangular ramp so people can travel from the ground level up on to the bridge level.'

'It just sounds incredible, [the idea that polystyrene] is strong enough to support a multi-lane highway with cars and trucks. What did you make of the idea when it was first suggested?' Jim asked.

'At first, I thought it was nuts!' Ailie said. But the more she looked into it, the more sensible it began to seem.

Expanded polystyrene (EPS) has a very high compressive strength. It's very durable. And it's wonderfully lightweight. Putting polystyrene blocks in place is a lot quicker and cheaper than moving great mounds of earth around, 'so it was much more cost effective.' And this lightweight fill had another major advantage. Piling earth up on the surface to create off–on ramps for bridges often causes the ground below to become unstable. Much of downtown Boston is built on landfill, so this was likely to be a problem. (Building bridges on dense clay or solid bedrock is much easier.) If heavy piles of earth were replaced by lightweight polystyrene, the downward pressure on the ground would be dramatically reduced, simply because polystyrene is so much lighter. It means 'you can build ramps without having to worry so much about the ground conditions below the ramp.' This innovation knocked months off the schedule and saved a lot of money.

'It was a great part of the project, actually,' Ailie said, clearly pleased to have found a new, improved solution to an age-old problem. 'It was great to innovate in this way.'[3]

Expanded polystyrene has since been used in construction projects 'all over North America'. And 'There's lots in northern Europe too,' Ailie said. 'Norway and Sweden do it a lot.' It's a good solution when the ground is frozen for much of the year. 'Moving earth around when it's very cold is very difficult, whereas with expanded polystyrene you just move the block in.'

---

3 The Boston Big Dig was the first construction project in the US to install blocks of expanded polystyrene as a lightweight fill but not in the world. The Norwegians were the pioneers, first using it in 1972, to shore up a bridge ramp that was sinking into the ground at a rate of 30 centimetres a year.

After seven years in the USA, Ailie returned to the UK to manage the transformation of London's St Pancras station into a terminal that was fit for Eurostar. 'It was transforming a one-hundred-and-fifty-year-old building, which was in bad repair and had been neglected, certainly since World War II, and turning it into what you see now ... The Eurostar trains were twice as long as the trains that St Pancras was built for. The old train station is about two hundred and thirty metres long and the Eurostar trains are four hundred metres long, so we had to extend the old part of the station, as well as making sure that the one-hundred-and-fifty-year-old building was structurally competent and safe to take a completely different type of train ...

'We needed to find room for everything that goes with an international train station: the fire safety systems, the signalling systems, the air conditioning, the gas, the electricity, the communication systems, the access control. We had to find nooks and crannies that we could use in that old station to fit in all of the new technology that comes with modern railway stations.'

'Was there anything in particular that made you think, "How on earth are we going to do this?"' Jim asked.

Ailie laughed. Her thoughts on one occasion were 'a lot stronger' than politely wondering 'How on earth ... ?'

The entire ground level of St Pancras station – the platforms, the track, the trains, the passengers, the floor itself – is supported by a subterranean 'forest of cast-iron columns'. 'One of the beauties of cast iron is that it will take a lot of vertical load,' said Ailie. If you push down on it, it will cope. But it doesn't take lateral load very well: 'If you push sideways it will snap like a carrot.'

The transformation of St Pancras station was all about creating a light and airy undercroft area beneath the platform to accommodate all the shops and restaurants. In order to get light flooding down into the lower level, the plan was to cut out parts of the existing platform, creating light wells the size of swimming pools. Engineers proceed with caution, measuring everything as they go, constantly checking that what is actually happening corresponds with what they predicted would happen. The first bit of the platform was cut out and the impact was duly measured. 'And we found that [the cast-iron pillars] couldn't take the lateral load,' Ailie said. 'It was larger than expected.' The columns that were supporting the entire platform, and the roof above it, couldn't take the new lateral loads that had been introduced at platform level when pieces were removed, and they were in danger of snapping like carrots if the work continued as planned.

Under normal circumstances, these columns could have been replaced with concrete struts capable of withstanding lateral and vertical loads. Part of the point of the project, however, was to preserve the historic character of this Grade 1 listed building. And Ailie was committed to using 'as much of that one-hundred-and-fifty-year-old architecture and structure as we could'.

For about five seconds, Ailie thought, '"It would be easier to demolish this lot and start again." Then it was a case of: "No. We've got to find a way."'

'The solution was to separate the platform level from the cast-iron columns,' she explained. So they introduced a pot bearing (similar to the kind of bearings you see on bridges) between each vertical cast-iron column and the deck that was holding the trains up. 'The bearing . . .

acts like a piece of soap. It allows the deck to move but prevents the lateral load from being translated into the cast-iron columns.'

'Did you immediately think, "Yes, that'll do it!" Or did it take a long time to find a solution?' Jim asked.

'We knew that a bearing of some sort was going to be required. We came to that fairly quickly.' But working out 'all of the details around it' took time: 'What type of bearing? How were we going to construct it? We needed to put a grout layer in between the bearing and the deck, and that had to be really high strength [to absorb the horizontal load].'

'I'll bet!' said Jim. 'You're supporting a platform with very long trains. And one of the largest roofs in the world . . .'

'It wasn't a normal use for a pot bearing,' Ailie conceded. And the devil was in the detail. Designing a bearing that could withstand loads on this scale involved a lot of 'really clever engineering which took weeks to work through'. They needed to work fast to avoid falling behind schedule, and the acceptable margin for error was small. Minor design flaws can have major consequences when they are reproduced 680 times.

The new St Pancras international station opened in 2007, accommodating not only the long Eurostar trains but shops, champagne bars and restaurants. 'It was fantastic!' Ailie said. 'Even now, when I go into St Pancras station I get a tingle down the back of my

> 'Even now, when I go into St Pancras station I get a tingle down the back of my neck'

neck. And it's not just me that feels like that. Thousands of people that worked on that station will have the same thing. That's why building infrastructure is so great. I can tell my kids what I did. And they can tell their kids.

'I was honoured to be part of the welcoming party for the Queen when she came to open up St Pancras station. That was a very proud moment. She obviously talks to hundreds of people every day. But when [the Queen] was talking to me about St Pancras station, I felt that I was the only person in the world she was interested in. She's a class act . . . I was expecting her to say something about me being a woman, but she said: "Ah. What a fantastic part of your career. What was the biggest challenge?" And I loved that.'

Next up was Crossrail, the new east–west Queen Elizabeth line for London Underground. Ailie worked on the challenging Central London section, managing a huge team that was responsible for constructing 42 kilometres of tunnels[4] and seven new stations under London, while keeping the city moving at the same time.

'At peak, about six hundred and sixty people were in my team to design and construct Crossrail . . . I tell you, that project is going to transform the way London Underground operates. It will add ten per cent to London's capacity, just like that!'

'Thirty per cent of engineers in your top team were women,' Jim said, impressed. 'Was that the result of a deliberate strategy to recruit more female engineers?'

'Thirty per cent of Bechtel's engineers on that project

---

4 With advice from, among others, Robert Mair. See p. 171.

were women. Obviously, thirty per cent isn't good enough. It should be fifty per cent. But it's a lot better than the industry average, which I think is in single figures.' (Currently about 8 per cent.) 'So it was definitely purposeful. When you've got such a big infrastructure project, you've got the opportunity and responsibility to invest those public funds in a way that lifts the sector up. I do think it's important that customers like TFL, Network Rail and HS2 understand, which they do, the opportunity and responsibility they have for putting programmes in place that increase the number of women who work on these jobs. We could make very purposeful decisions about who we hired; who we promoted into various positions; and who we asked to be a project manager. And we did that in a way that enabled a more diverse team to grow.'

'How easy was it for you to find the right people to fill these jobs?' Jim asked.

'That varied. If you look really hard, say, at the start of one of these processes, you will always find women (and members of other under-represented groups, actually) who, for a whole number of reasons, a lot of it being unconscious bias, won't have been promoted when maybe they should have been. There are changes you can make to increase the diversity of a team, by promoting from within. In terms of hiring new people, it depends on what stage of people's career you're looking to hire in. Coming out of university, the diversity is improving every day. The diversity of candidates who apply to the graduate programme is quite good. But then the industry is tending to lose some of the women. So

'There are changes you can make to increase the diversity of a team, by promoting from within' that's what Bechtel is looking at really hard. What's causing women, or people from diverse backgrounds, to leave a company?

'What I found really useful for me was recognising my own unconscious bias. I was hiring and I was talking to a head-hunter at the time. The head-hunter was saying: "What are you really looking for?" So I leant back and closed my eyes and I was just shocked to find that, in my mind, I was thinking of a white middle-aged man. All the people I had seen operating successfully in that role had been a white middle-aged man. So that's my bias. I've been working in this industry for thirty years. And I thought, "Geez, if I'm thinking that, it's no wonder we've got a problem!"

'The first thing is to be aware of our [unconscious bias], and the second thing is to do something about it. And that's what we're doing.

'The projects I've worked on have been projects that have changed cities. And I think that's what makes me even more determined to help boys and girls, but particularly girls, just to be aware of the type of career they could have in engineering. It's not all about standing on an oil rig with a hard hat and a messy old orange vest.'

'It's not all about standing on an oil rig with a hard hat and a messy old orange vest'

MOLLY STEVENS

*'We're designing materials to help
the body repair itself'*

Grew up in: France
Home life: married with three children
Occupation: bioengineer
Job title: Professor of Biomedical Materials and Regenerative Medicine, Imperial College London
Inspiration: seeing a picture of a boy with liver failure
Passion: inventing new bio-materials to help repair our bodies
Mission: to find inexpensive solutions to global health problems
Favourite invention: an injectable gel that encourages new bone to grow
Advice to young scientists: 'Surround yourself with good team players'
Date of broadcast: 22 November 2011

Professor Molly Stevens is a bioengineer. She invents new materials and techniques that help our bodies to heal themselves by growing new body parts: bone, cartilage and heart muscle, for example. Having studied pharmacy at Bath University, she did a PhD in single-molecule biophysics, in part because it sounded like 'the hardest thing possible'. Struck by an image of a small boy who was suffering from liver failure, she attended a talk by the founding father of tissue engineering and decided to use her scientific knowledge to tackle issues in human healthcare. She co-invented an injectable gel that tricks the body into making more bone and a material to replace joints. If all goes well, her bone-repair gel could improve the lives of millions of people with seriously broken bones or severe osteoporosis.

Jim Al-Khalili interviewed Molly in 2011 after pre-clinical tests had delivered some very encouraging results.

Molly Stevens grows bones. Not so long ago, the idea of growing body parts was a pipe dream: the stuff of science fiction, not reality. '[Tissue engineering] is moving really, really fast,' Molly told Jim.

The human body has an extraordinary ability to repair itself. Skin forms scabs over wounds, and broken bones grow back together after a small break or scrape, but some fractures are just too big, and some tissue too damaged for our bodies to fix themselves. And so, scientists around the world have been trying to create synthetic tissue to help heal patients with more severe needs.

Engineers working at MIT made synthetic skin (inventing a polymer that acted as a scaffold on to which new skin could grow). It was first used to treat patients with severe burns in 1979. But not all body parts, however, regenerate as readily as skin. In 2008, Doris Taylor, a medical researcher at the Texas Heart Institute, took a heart from a dead rat and washed the cells away, exposing a three-dimensional structure made from collagen and other large molecules. This scaffold, or phantom heart, was then populated with heart cells taken from a new-born rat, which divided and multiplied and, at one magical moment, started to beat. News of this breakthrough spread across the world, together with an image of a pumping rat's heart suspended in a glass jar. Our imagination ran wild. There were visions of lab-grown human hearts, beating in jars, lined up in hospital, ready to replace damaged or diseased organs.

'How far off are we from that [human] heart in a jar?' Jim asked.

'That is actually so difficult to do!' Molly laughed. 'It's really, really hard. The thing to do really is to get your expectations a little bit more realistic, and then target some slightly lower-hanging fruit.' Growing heart muscle, for example, to help rebuild the heart wall if it's failing after a heart attack. Or trying to reconstruct the valves in the heart. 'That is more the way the field has gone so far.'[1]

> 'There'll be people working in this field on pretty much any tissue you can think of'

Bio-artificial blood vessels would be incredibly useful, but creating a material that can fight infection and pump blood is quite a challenge. Scientists have managed to grow liver and kidney tissue. Soon 'there'll be people working in this field on pretty much any tissue you can think of,' Molly said. Growing transplantable organs is the holy grail of tissue engineering, and 'certainly' there are still people who are 'aiming for the whole heart.' But 'this is still further from the clinic. It's still experimental at the minute.' But it is exciting, nonetheless.

'You can actually see the cells maturing and forming tissue. And you'll see those human heart cells coming together, and it will start to beat! . . . It's pretty amazing when you see the cells in the gel actually starting to beat together in synchrony.' Witnessing a pulse emerging from a group of cells, 'that's pretty cool.'

It might also make some people feel as if scientists have created life. 'Have there been accusations about being Dr Frankenstein or playing God?' Jim asked.

---

1 Since Molly's interview with Jim in 2011, her group has developed several new materials for regeneration of the heart.

'For me?' Molly replied, surprised.

'For you, or levelled against people in this field. Is there still controversy?'

'Um . . . I don't know. I don't know. I think what we're doing is fairly straightforward. We're designing materials to help the body repair itself or to detect disease, and there's not really anything Frankenstein about that.'

}

'I always keep an eye on how what we do can be useful,' Molly said. But she has always been driven by curiosity.

'I get the impression you've *always* been interested in everything,' said Jim.

Choosing which A levels to study was a 'nightmare' for Molly. In the end she picked sciences and French, making a promise to herself that she'd keep up her interest in geography and history as hobbies. For many, pharmacy is a vocational degree but Molly had no intention of becoming a pharmacist. She chose to do a degree in pharmacy because of all the different kinds of exciting things she could learn: 'everything from human biology to chemistry to designing drugs to pharmacology'. And because she thought spending five years at university studying medicine was 'far too long'. Having completed her three-year bachelor's degree and travelled in South-East Asia, however, she then embarked on a three-year PhD in 'the hardest possible thing': single-molecule biophysics.

A degree in pharmacy was not the best preparation for such a challenging new subject, but things worked out rather neatly. Molly got a job as a locum pharmacist, which helped to pay the bills while she was studying for her biophysics PhD. With the calculations for her PhD safely tucked away under the counter, she could whip them out whenever the pharmacy was quiet and get calculating. 'It saved me lots of time!' she said.

Single-molecule biophysics (now more commonly described as nanoscience) involves looking at how life works right down at the level of individual molecules. 'Essentially I was taking peptides, which are small molecules that had self-assembled together, and pulling them apart to figure out what the forces were between them.' The calculations she made of the forces acting between these molecules were some of the first ever measurements to be made at this scale. It was, she said, a 'pretty exciting time', but something even more exciting – and entirely unexpected – happened next.

In the year 2000, Molly flew to San Francisco to give 'a twelve-minute PhD talk in biophysics' at one of the many *ad hoc* millennial science conferences that were popping up all over the place that year. She had never had much money and was not at all convinced that it was worth the air fare, but her supervisor encouraged her to go. 'I remember being terrified!' she said. Her talk went just fine, but she had other worries on her mind. Namely, she was 'absolutely unsure what would come up next' in her life. 'Then I happened to be walking past another room that was not at all related to my talk, and I saw a picture on the screen of this little boy with really, really bad liver failure. It sort of grabbed me and my attention.'

'I was absolutely unsure what would come up next'

She stopped to watch the presentation: a talk by Robert Langer, a world-famous MIT scientist and the founding father of tissue engineering, about his work trying to grow new liver tissue. (Given the liver's remarkable regenerative properties in our bodies, it had seemed as though it might be one of the easier organs to grow. It wasn't. When you consider that in just 1 gram of liver there are about 100 million cells, all of which need to be aligned in precisely the right way, it's not hard to understand why it's difficult to grow new liver tissue. Just because nature can do something, it doesn't mean we can.)

It was 'absolutely amazing', Molly thought, that people could design materials that would help the body to heal itself, and, 'within a few minutes', she knew exactly what she wanted to do for the next stage of her life. 'It was a very marked event' and a huge relief. 'Because, obviously, I do like to have something to focus on,' she said. 'It totally, totally inspired me ... to switch fields; to keep the biophysics skills I had acquired and apply them in a completely different way.'

Jim couldn't quite believe that Molly had had the confidence to ask Robert straight after the talk if he would take her on as a researcher. 'There

> *'I do like to have something to focus on'*

was Bob Langer, father of tissue engineering, and there were you, with no experience in the field at all,' he marvelled. Amazingly, Robert Langer said yes. 'Were you surprised that he took you on?' Jim asked.

Molly hesitated for a few seconds. 'No, no ... not really,' she admitted, 'though I suppose I really should have been!'

Jim admired her bravery. Molly said it was more likely her being 'foolhardy', and that she had strong support from her PhD supervisors, which helped.

She joined Langer's lab at MIT almost immediately and, inspired by Bob's work on regenerating livers, she soon turned her mind to bones. Bone tissue, like liver and skin tissue, has the ability to repair itself and was thought to be another promising area to research.

If we fracture a bone, new tissue grows to close the gap, providing it is not too big. Bones are able to do this in part courtesy of a thin layer of stem cells – cells that have the capacity to multiply rapidly – that live just underneath their outer surface. An injury to the bone 'wakes up' these stem cells, prompting them to proliferate. New bone tissue grows and the gap is closed.

But for millions of patients nature's healing is not enough, and a bone graft is needed – to treat a badly broken arm, for example. The classic way of doing this was for a surgeon to carve off a bit of bone from the patient's hip and insert it elsewhere in the body. The operation was normally successful but would often leave patients with painful hips for up to two years.

If tissue engineers could create new bone tissue, then perhaps this pain could be avoided?

The traditional approach to tissue engineering at the time was to grow tissue in Petri dishes, but Molly and her co-workers believed the best place to grow new bone was probably in our bodies. Instead of extracting cells from the patient, and growing them in the lab (perhaps on an artificial scaffold like the rat's heart), she decided to investigate the possibility of encouraging human bodies to grow more bone inside themselves. The front calf bone (the tibia) seemed like a good place to start with. It often got damaged and seemed to heal quite well. There were plenty of stem cells underneath the outer layer and a relatively large surface area on which new bone could grow. Could this bone be fooled into thinking it needed to repair itself, so as to make new bone that could be harvested and used to fix a problem elsewhere? Could the stem cells hidden underneath the surface of the tibia be activated in the absence of an external injury?

Careful observation of how bones heal themselves gave the team hope that if they were to force a gap between the outermost layer of the bone and the bone itself, it might encourage the trapped stem cells to start multiplying to fill the gap. Molly set about trying to invent a material that could be injected just under the surface of the bone and would somehow force these two tightly packed layers apart. Using her knowledge of the microscopic structure of materials, she invented a gel that would be tolerated by the human body and would solidify within 30 seconds of being injected into the bone.

In this way, she hoped to coax the body into a regenerative process, 'essentially fooling it into thinking it has to do a wound healing-type response and form a massive amount of new tissue' which could then be harvested and used to repair bone damage in areas of the body that were unable to heal themselves.

In theory it was an elegant solution: harnessing the innate smartness of our bodies to get nature to do what we wanted. But often, when scientists think they understand how nature works, they are proved wrong, so while Molly was excited by what her team might be able to achieve with her cleverly designed gel, she, nonetheless, rated her chances of success at 'about fifty-five per cent'.

Undaunted, Molly carried on, because the prize was worth it. And, after a couple of years of trying to get this idea to work, one day she was rewarded by a 'massive amount of new tissue'. 'That was a huge, huge eureka moment,' she remembered, laughing. 'We were screaming all over the place! We were all very, very excited.' Keen to see for themselves, all the members of the team rushed to look down the microscope, jostling for position. The sheer quality of the new bone was a revelation. The organisation of it, 'right down through all levels of the tissue', was a perfect replica of the bone to which it was attached. 'It was just perfect native tissue,' Molly said.

Bone tissue is much more varied and much more 'alive' than you might imagine. 'They're not the dry old things most of us imagine hanging on a classroom skeleton. They are packed with tiny blood vessels' and have these large and distinctive circular features called haversian systems. 'It's a very peculiar-looking organisation,' Molly explained, and it's very difficult to engineer. 'This new tissue even had those in it!' she exclaimed. 'You just don't normally get that because these systems are so complex and so organised.' Previous attempts to grow bone had resulted in new tissue that was a pale imitation of the real thing. But in Molly's experiment, the two areas of bone – pre-existing and recently grown – were indistinguishable. Furthermore, mechanical testing confirmed that the new bone was just as strong as the original.

'How long did you stay on this high?' Jim asked.

'Well,' laughed Molly, 'I'm sort of always on a high anyway!'

Molly has described what she does as 'geeky, hard-core science', but her main aim is to help people. It can be a long, hard slog to move from a eureka moment in the laboratory to patients reaping the benefits, and there is no guarantee of success.

Molly, however, is determined to make use of break-throughs in understanding the fundamental science to drive innovation in medicine and develop new treatments. And by keeping her eye firmly on the end goal, her outcome-oriented approach to tissue engineering has 'attracted more funding than any young scientist I know', according to a distinguished colleague. She also attracted some attention from the media, as a leading innovator under forty and 'the woman who grows bones'. Very reluctantly she posed for a photo shoot for *Vogue*, on condition that the accompanying article would be strictly science-based.

Molly's understanding of fundamental science teamed with a practical determination to use knowledge to create treatments is a powerful combination. 'I'm really, really keen for what we work on to be able to make it to patients,' she said. 'I think that's really, really important to me.' Her eye is always on how the science that she and her colleagues work on might translate into something that is medically useful. 'I don't want to spend a lot of research money and a lot of time making things that are not, in the end, going to be useful for people.'

'I'm really, really keen for what we work on to be able to make it to patients'

The best way to achieve this, she thinks, is to include everyone in the research process right from the start. All too often, the end user – the surgeon – is the last person to know about a new technology that has been carefully designed to help them. Molly involved surgeons in the research process from day one, rather than asking for their comments at the end, when most of the research money would have been spent. At her wonderfully multidisciplinary lab at Imperial College, engineers, biologists, physicists, material scientists, surgeons and mathematicians work together. 'It's amazing,' Molly said. 'Group meetings can be incredibly innovative. [You] bounce a lot of ideas around and also get a healthy dose of realism. We might propose something from an engineering viewpoint, and the surgeon might turn around and say, "Hang on a minute, we don't do that in practice."'

Jim wondered if consulting such a wide range of experts led to friction among the scientists, each with very different concerns and agendas. 'No, no!' Molly assured him. 'It might save you a couple of years of useless research, so it's very, very useful.' Surgeons are able to articulate their needs and so set research objectives that could bring about real benefits to patients. It's 'extremely valuable both ways'. In multidisciplinary science, the total expertise is greater than the sum of its parts.

'Even if it doesn't work quite the way you thought it would, something interesting always comes out of it'

'The research has gone much better than I could possibly have hoped for,' Molly said. 'And even if it doesn't work quite the way you thought it would, in science something interesting always comes out of it.'

The promise of saving a little boy with liver failure inspired Molly to enter the world of tissue engineering. That image and a desire to help patients remains with her. She is acutely aware of 'the massive need for new technologies and for cheap medical devices in global health-type situations', and it drives her forward. 'A lot of our innovations remain just in the Western world,' she said. That needs to change.

Since recording this interview in 2011, Molly has stayed true to her word. Her team has made major advances in the development of medical technologies for use in countries where access to healthcare is poor. They have invented diagnostic tools that allow patients to be tested for illness or disease wherever they live and sent by mobile phone to medical professionals elsewhere. One invention has been used successfully in Uganda to monitor the blood of Ebola survivors. Another is being tested in South Africa for early detection of HIV infection.

ACKNOWLEDGEMENTS

When Gwyneth Williams (controller of BBC Radio 4, 2011–19) commissioned 'a conversation between scientists' to be broadcast on prime-time BBC Radio 4, some feared that only a handful of scientists would be able to talk in an engaging way for half an hour. Thank you, Gwyn, for having the idea and for creating the space for us to prove them wrong. *The Life Scientific* was, as I have said before, a brilliant and bold commission. Alan Samson at Weidenfeld & Nicolson made another bold and brilliant move when he decided to publish not one, but six books of *The Life Scientific*, arguing that one book would not do justice to all fascinating and admirable scientists who have been interviewed on the programme. Thank you, Alan. You were right, of course.

Mohit Bakaya (who is now the new controller of Radio 4) believed wholeheartedly in the idea. Thank you, Mohit, for not laughing when I suggested 'maybe it could be a bit like *This Is Your Life* for scientists?' Thank you too for appreciating the ingenuity and creativity involved in scientific experiments. The editor of the BBC Radio Science Unit, Deborah Cohen, never doubted that we would make it work. Thank you, Deborah, for all the edifying conversations we have had over the years and, in particular, for your unflinching commitment to maintaining a

50–50 gender balance across the strand.

I feed off the knowledge and enthusiasm of all my talented colleagues in the Radio Science Unit daily and special thanks are due to Beth Eastwood (who produced the interview with Christofer Toumazou), Geraldine Fitzgerald (Tony Ryan), Melissa Hogenboom (Jackie Akhavan), Michelle Martin (Mark Lythgoe) and Pam Rutherford (Robert Mair); and to Maria Simons for whom nothing is ever too much trouble. And thank you, of course, to Jim Al-Khalili for all the wonderful interviews. I remain in awe of your ability to absorb information at speed and to remain calm. And I greatly appreciate your gentle and intelligent approach. It is an honour and a privilege to work with you.

These stories belong, of course, to the scientists who were kind enough to agree to be interviewed by Jim and who were kind enough to talk to me, often at great length! Thank you all for being so fantastically generous with your time and for being so honest; for sharing the lows as well as the highs. My gifted niece, Daisy Buckley, tackled several chapters. Thank you, Daisy, for your glorious prose and for lightening my load. Paul Murphy at Orion skilfully got me started with the whole business of writing a book. The text benefitted greatly from some astute editing by Celia Hayley. And I absolutely love James Jones's wonderful cover design.

So many friends spurred me on and listened patiently while I said the same thing over and over again. Thank you to you all. And thank you to my mother, Celia, brother, Will, and sister, Camilla, for your unfaltering support. I started to write these books when our teenage daughters, Eliza and Rosie Quint, started to develop interests of their own. Thank you, Eliza, for your kind words of encouragement and some excellent editorial advice. Thank you, Rosie, for spotting where I was going wrong, and being an enthusiastic ambassador for *The Life Scientific: Explorers* and for your insightful comments. Inevitably writing this book (on top of everything else) has gobbled up vast swathes of time in which I might otherwise have been doing other things. Thank you, Mike, for your patience with the process and, more importantly, for your love.